Discovering Science Series

Physical Science

by Connie Blood

illustrated by Carol Kotsher Marden

FS-10138 Discovering Science Series–Physical Science

23740 Hawthorne Blvd.
Torrance, CA 90505

Table of Contents

Introduction

Science is the observation, investigation, identification, description, and theoretical explanation of natural phenomena. It provides us with the ability and methodology to understand what is happening in the natural world around us. By understanding more about natural phenomena, we gain a better understanding about why events occur and we are better able to either control or adjust to our natural world.

Physical science is a subject that is fascinating, exciting, and most of all, fun to learn. For the learning process to occur, science must be introduced and taught in a manner that creates curiosity and excitement. Too often students are exposed to science through uninspiring texts and learn by memorizing facts and answering questions. This methodology requires little preparation time on the part of instructors and offers little motivation to generate interest on the part of students. These techniques produce canned lectures, boring demonstrations, and laboratory exercises that do not provoke questions from eager and inquisitive learners.

Provocative and meaningful science instruction should emphasize the thought process and not simply the content. Learning how to observe, organize, make predictions and inferences, draw conclusions, and communicate results are the ways students "LEARN HOW TO LEARN."

The learning process should be guided by thought-provoking, student-generated questions. The teacher's role is that of a facilitator of learning experiences that help students create their own questions and investigations. Understanding beforehand what information students already know helps establish a starting point for exploration and discovery. Manipulation, exploration, and experimentation are the most crucial elements involved in the "learning of science." Hands-on activities and sufficient time to examine and investigate ideas are a priority.

To be good problem solvers, students need to be persistent and flexible. They must learn not to jump to conclusions or simply guess at answers. By being flexible they will be able to change direction when needed. By being persistent, they will not give up easily or turn to the teacher for a solution.

It is with these ideas in mind that this book of physical science activities has been produced. The activities are, hopefully, both fun and challenging. They have been designed to ensure that a great deal of advanced planning is not required. Whenever possible, attempts have been made to stretch the investigative possibilities by suggesting additional activities students can explore. Wherever appropriate, graphs and charts are available for recording data.

Sections for the teacher provide brief background information on each topic. In addition, suggestions are given for lessons, demonstrations, and discussion questions. Teacher pages refer to student activity pages that correlate with each topic. Student activity pages follow teacher pages in each chapter.

SCIENCE AND COOPERATIVE LEARNING

Twenty years of research demonstrates that when cooperative grouping is well implemented, learning is increased. There needs to be a balance between individual and cooperative learning but it is important for students to have the experience of working in groups. The skills learned in group situations—interaction, sharing, cooperation, and trust—serve us well in all of life's situations and are of particular value in the workplace. Industry in particular recruits those individuals who are able to combine technical skills and the ability to work well with others.

Following are some ideas for using cooperative grouping in your science classroom.

BRIGHT IDEAS FOR TEAM-BUILDING ACTIVITIES

1. Activity Sheets
   Give each student in the group an activity sheet. Have students discuss questions and answers together and have each person record the answer on his or her own sheet. Let students know ahead of time that only one paper from each group will be graded and the others in the group will receive the same grade. This encourages everyone in the group to cooperate and gives the low achiever a chance to excel.

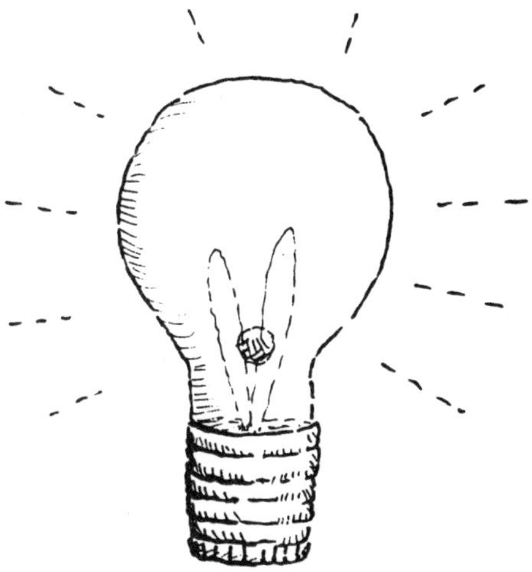

2. Have groups practice thinking skills by giving them a photo of a science-related topic such as pollution. Have students discuss what may have caused the situation in the picture and suggest ways it can be improved. Answers should be shared with the class.
3. Ask students to predict what might happen in an experiment before it is carried out. Share group predictions with the class.
4. Give each group a copy of a science-related article to read and discuss. Have students find out what the problem is, what caused it, and how they think the problem can be solved. Share answers with the class.
5. Have groups contribute to a bulletin board. For example: When studying the digestive system, assign each group a different organ and have the group research, write, and share information about its organ and the organ's part in digestion. Use each group's writeup as part of the bulletin board, placing each writeup on the appropriate organ on a large diagram of the digestive system.

GUIDELINES FOR SETTING UP COOPERATIVE SCIENCE GROUPS

1. In a group of four, select one high-ability student, two average-ability students, and one low-ability student.
2. Let students pick a team name and create a sign with the team name.
3. The team reward system should be kept separate from grades. Even though students are working together to learn and study materials, keep individual accountability for grading.
4. Have a record-keeping system that is visible to students so they can watch their progress. A large wall chart with team signs works well. Try to keep competition close. Award points for as many criteria as you can. Some suggestions are:
   group cooperation and behavior
   group or individual improvement
   homework completed
   participation
5. Let students make a list of rewards that the team with the highest points can earn. Points should be added up at the end of every two to four weeks. At this time students should also get into new groups.
6. Assign jobs to the members of each group as follows:

JOBS

Director
- Reads all directions out loud to group
- May ask teacher questions for group
- Leads group discussions

Recorder and Reporter
- Records all observations and data from experiments
- Reports results of group work to class

Materials Manager
- Picks up all material for experiments
- Returns materials when experiments are completed

Experimenter
- Performs the experiments for the group
- Helps materials manager clean up work area

   Job assignments can rotate for each new task. Badges can be made with the job titles written on them. Laminate the badges and let students wear the appropriate ones.

7. To call on students for answers, number the students in each group one to four. The teacher can call a number at random. If 2s are called, then the #2 students in each group will supply the answer. A spinning wheel, a die, or playing cards can be used to determine who will answer the question.

   All members of the group need to discuss and know the answer to each question because the teacher may call on any member.

Electricity

CHAPTER 1

Electricity Kit
Student Letter
Page 7

TEACHER NOTES

The following activities revolve around the use of an electricity kit which contains materials brought by students from home. Students who do not have kits can be grouped with students who bring kits and can act as helpers. All of the activities can be done in small groups of two to three students. The activities done with the kit build on one another and should be done in the order presented.

Student activity sheets can be completed individually and then discussed in the group or the group can complete one activity sheet together and share results with the class.

Have students assemble an individual ELECTRICITY KIT consisting of the following materials:

- 1 six-volt battery
- 1 small porcelain socket
- 2 flashlight bulbs that screw into the socket
- 6 feet of thin copper wire
- Small and large nails
- Aluminum foil
- Piece of wood (approx. 6" x 10") to mount socket and battery
- Hammer, small screwdriver, and wire cutters

See sample letter to send home with students.

In preparation for activities, have students mount the porcelain socket on the board with screws that come with the socket. The battery can be held in place with four large nails. Hammer a nail on each side of the battery as shown below.

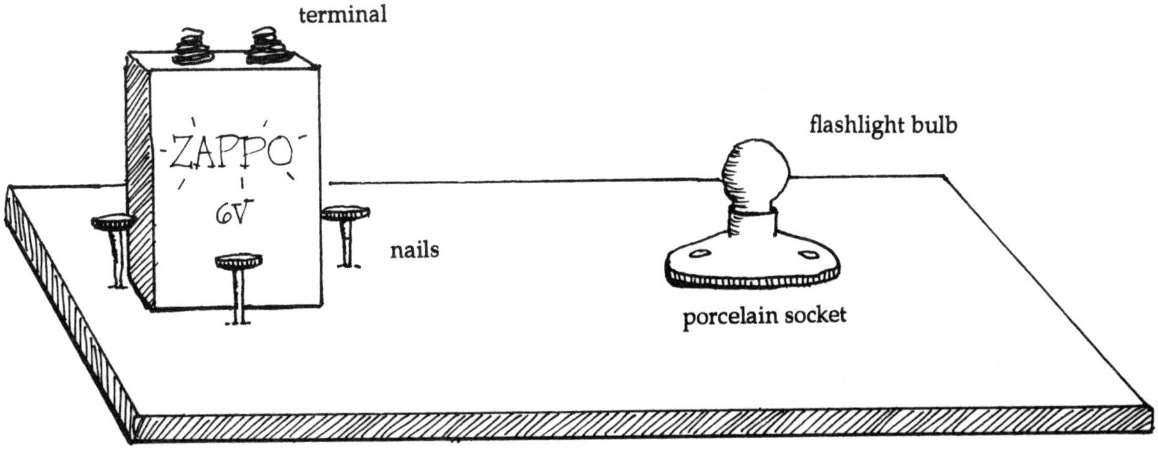

Circuits
Pages 8 and 9

Have students set up kits as shown on the previous page. Each student should then cut and strip two pieces of copper wire, each about 8" long.

QUESTION:

Using two pieces of wire, how can you get the bulb to light? (Give students plenty of time to experiment.)

ANSWER:

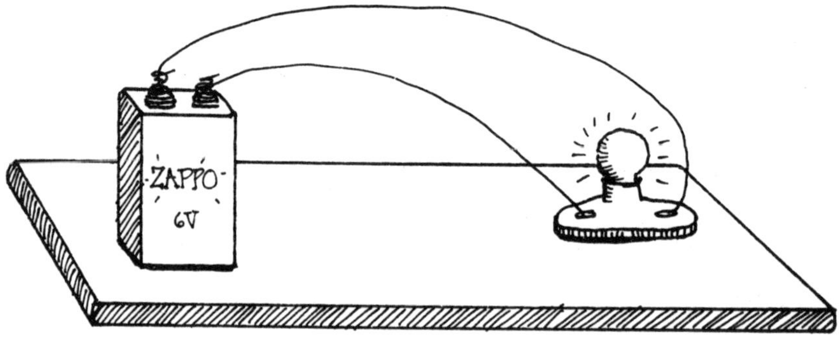

DISCUSSION QUESTIONS:

1. What happens when one of the wires is disconnected from the battery terminal? Why? (The light bulb goes off because the circuit is broken.)
2. Could you get the bulb to light using only one wire? (No)
3. Why does the bulb light up when all the wires are connected as above? (There is a complete circuit.)

Use the activity "Complete and Incomplete Circuits" to reinforce this lesson.

EXPLANATION:

Electrons travel in a path or circuit. When the circuit is complete, the electrons will flow and the bulb will light. When the circuit is broken or incomplete, the light bulb will not light because the electrons do not have a complete path on which to travel.

Electricity is the flow of electrons!

On the "Predicting Circuits" activity sheet students will predict which circuits will enable the light to go on. After students have made their predictions, supply some batteries, bulbs, and wires and have students test their predictions by experimenting! Discuss why some of the arrangements do not work.

Circuit Cards
Pages 10 and 11

Have each student make a circuit card as explained on the activity sheet. The circuit cards will help reinforce the idea that a complete circuit is needed to allow electricity to travel from one point to another. Students should use the answer sheet to record their findings. Since answers are on the back of each circuit card, students can self-check their answers for each card as they go along.

Electric Lingo
Parallel and Series Circuits
Pages 12 and 13

Electricians and scientists use special symbols to represent batteries, wire, bulbs, and circuits. The "Electric Lingo" page will help students become familiar with these symbols and give them practice using the symbols to represent electrical circuits.

Circuits can be set up in two different ways—in series or in parallel. In a series circuit, the bulbs are connected in a row. In a parallel circuit the bulbs are connected so each bulb is on its own circuit. Using this activity sheet, students can experiment with each type of circuit. They will also be able to use "Electric Lingo" to draw their circuits.

Making a Switch

Assemble the switch shown below and then discuss it with the class.

MATERIALS: 2 nails
Paper clip

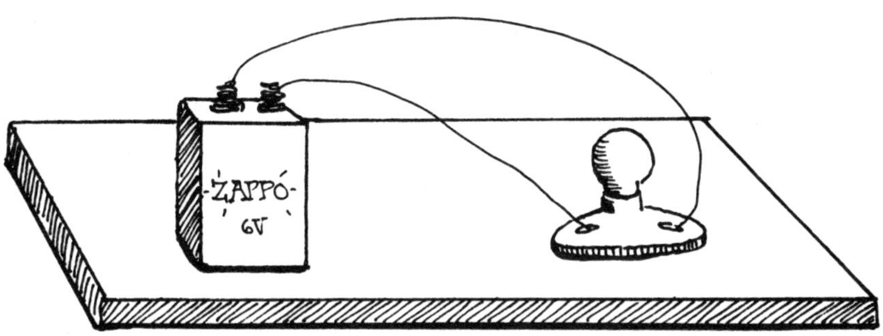

In the setup above, one wire must be disconnected from the terminal to turn off the light. Ask students if they can think of a way, using the two nails and the paper clip, to make a switch so the light can be turned on and off. You may need to give a few hints to get students started. It may be helpful to draw the setup above on the board while discussing how a switch can be made.

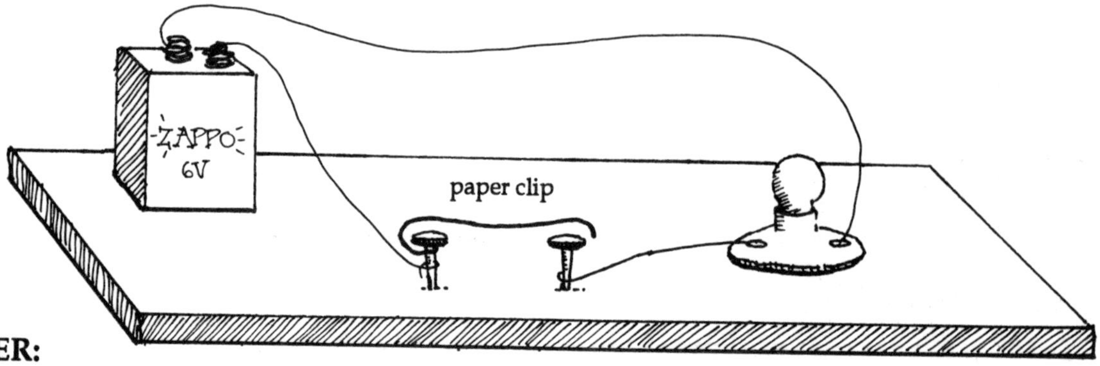

ANSWER:

Cut one wire halfway between the battery and the socket. Strip the wire where it was cut. Hammer two nails into the board fairly close together. Attach the stripped end of the cut wire to each nail. Bend the paper clip to fit across the opening between the nails. The paper clip will open and close the circuit and act as a switch.

DISCUSSION QUESTIONS:

1. Why doesn't the light go on when the switch is open? (incomplete circuit)
2. Why does the light go on when the switch is closed? (complete circuit)
3. Can you trace the flow of electrons through the circuit?

When switches are completed, turn off lights in the room and let students turn their lights on and off.

Short Circuit and Fuse

Electricity travels on a path or *circuit*. It will take the shortest path or the path of least resistance. In *a short circuit*, too much electricity is traveling on a short path and the wires become very hot. A *fuse* or thin wire in the circuit will heat up quickly and break, causing the current to shut off.

Make a fuse by setting up the following arrangement. Then discuss the process and the results with the class.

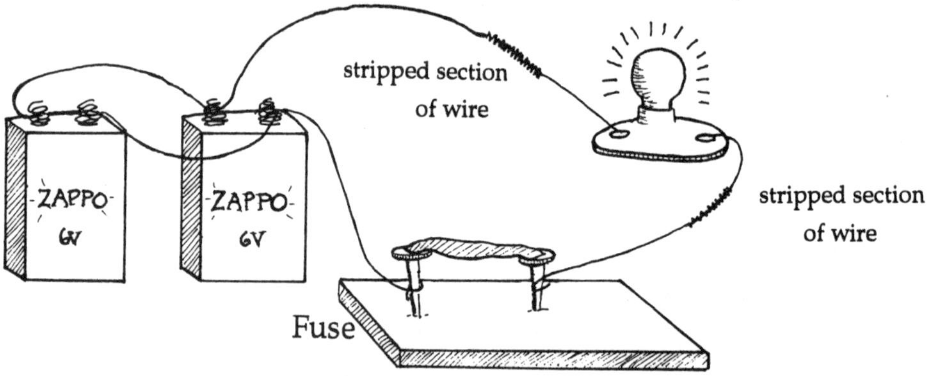

DIRECTIONS:

1. Place a very thin strip of aluminum foil across the two nails (in place of switch). The light bulb should be on when the foil is wrapped around the nails.
2. Touch together the two areas of stripped wire on either side of the bulb. This creates a short circuit and the foil (fuse) will heat up and break, cutting off the current. The light bulb will go off.

DISCUSSION QUESTIONS:

1. How do fuses and circuit breakers protect your home?
2. What happens if you add another battery to your circuit?
3. What happens if you make the foil strip wider?
4. Why do most new homes use circuit breakers instead of fuses?

Conductors and Nonconductors Page 14

Try placing different materials across the switch to see what materials will conduct electricity. If the material being tested will conduct electricity the light bulb will go on. See student sheet for suggested materials to test.

DISCUSSION QUESTION:

How do types of materials that are conductors compare to things that are attracted to magnets? (most are metals)

Electromagnets Pages 15 and 16

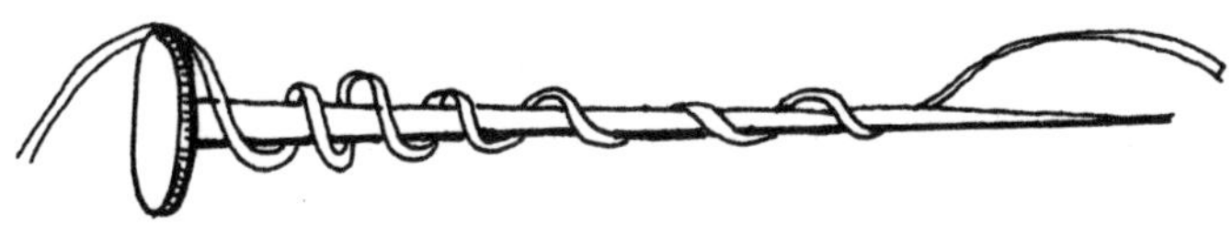

Have students wrap a two-foot long piece of copper wire around a large nail.

Attach one end of the wire to the terminal of a six-volt battery. Turn the electromagnet on and off by touching the loose end to the other terminal.

Activities to Try:

- Pass out iron filings and have students move the filings from one side of the paper to the other by turning the electromagnet on and off.
- Let students experiment to see what other things (such as paper clips, straight pins) the electromagnet will attract.
- Discuss the fact that electromagnets can be turned on and off. They can also be made stronger or weaker by increasing and decreasing the amount of current.
- Discuss ways electromagnets are used in machines. For example: Strong electromagnets can pick up a car and move it to a different place.
- Use the student graph pages to record observations from the following experiments: What happens to the strength of your electromagnet when
  - you add more batteries?
  - you make more turns of wire?
  - you change the size of the nail?

Let's Make a Battery
Page 17

Batteries make an electric current chemically. In a *wet cell,* two electrodes (copper and zinc) are dipped into an acid (*electrolyte*). A chemical reaction in the acid causes electrons to collect on the zinc electrode. The electrons flow through the wire to the copper, creating a current. In a *dry cell* (flashlight battery), the electrodes are zinc and the rod is carbon. The electrolyte is a paste instead of a liquid.

Students can take a D-cell battery apart in groups of two or three students. An automotive battery is filled with acid solution that is dangerous and should not be opened in the classroom. Also, alkaline batteries and lead acid batteries look like regular D-cell batteries but are not safe to open because of the chemicals.

To open a regular D-cell battery, follow these instructions:

1. Remove the outer covering carefully using wire strippers.
2. Remove the metal cap and take out the separator which is made of a tar-like substance.
3. Twist the carbon rod and pull it out.
4. Try to light a #48 light bulb with the black material left in the battery. Push the carbon rod slowly back into place and see where a #48 bulb will light and where a #41 bulb will light.

Have each student make a battery, using directions on the student sheet.

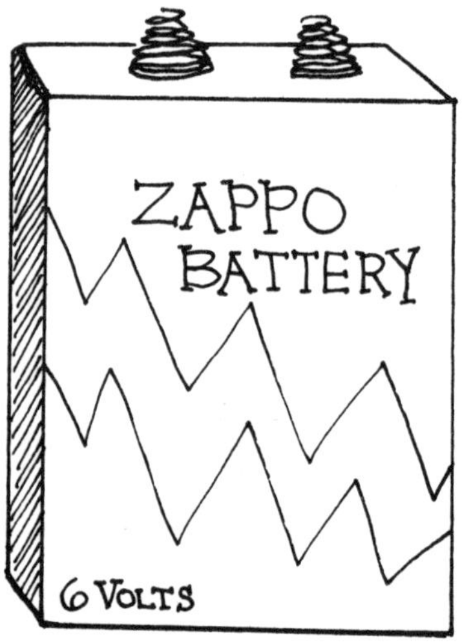

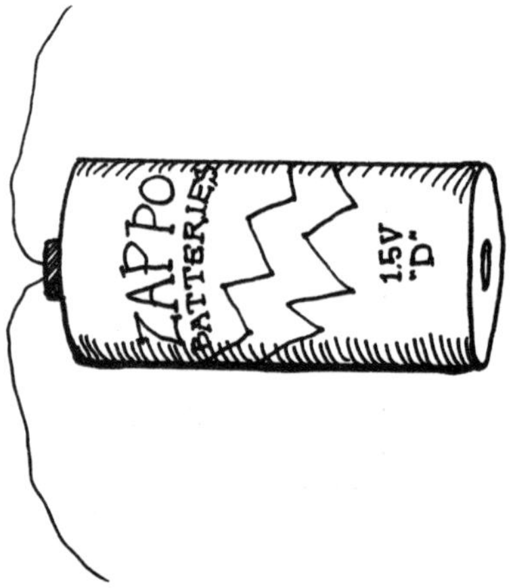

Dear Parents,

We are going to begin a unit in science on magnetism and electricity. Students will have the opportunity to do many hands-on experiments and often become so interested and excited about the activities that they want to take the equipment home. Although we have equipment at school for the students to share, we need to keep our materials at school for use another time. If your child would like to assemble his/her own kit for use at school and to take home after the unit is completed, here are the things that will be needed:

ELECTRICITY KIT

- 1 six-volt battery
- 1 porcelain socket to hold a flashlight bulb
- 1 flashlight bulb to screw into socket above
- hammer and small screwdriver
- several nails–small and large
- several feet of thin copper wire
- shoe box to hold supplies
- board (approx. 6" x 10") on which to mount the socket.

Name ________________________________

COMPLETE AND INCOMPLETE CIRCUITS

Using a colored marker or crayon, show where you would connect the wires to light the bulb.

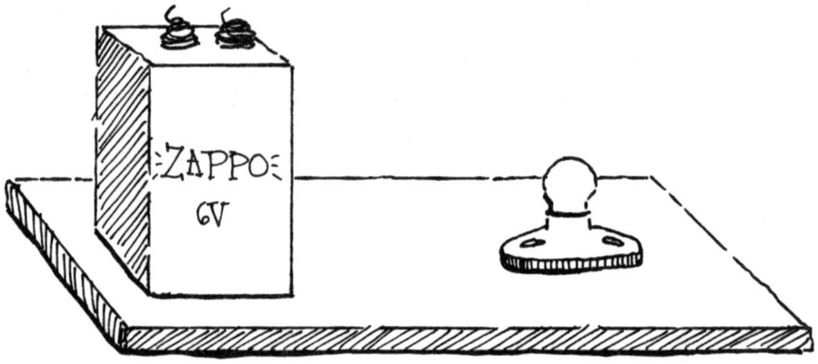

Using colored markers or crayons, show how you would change the wires in the picture above to make the light go off.

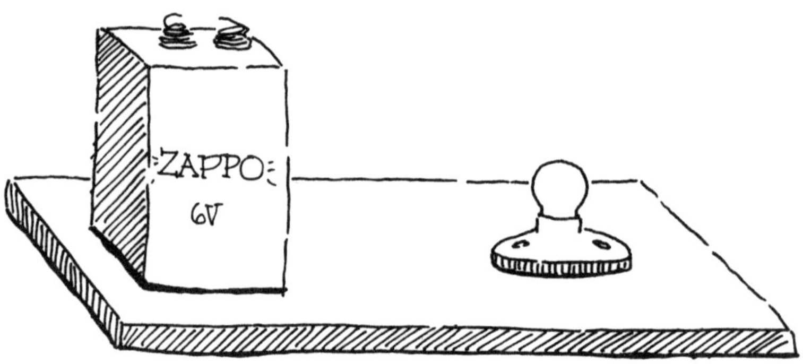

QUESTIONS:

1. What is a circuit?

____

____

2. What is the difference between a complete and an incomplete circuit?

____

____

3. What does a switch do?

____

____

4. Draw small arrows along the circuit in the first picture to show how the electrons flow through the circuit.

Name ______________________________

PREDICTING CIRCUITS

This is a D cell battery. The terminals are at each end. If you used one piece of wire and tried to get the bulb to light, which of the following bulbs would light? After you predict what will happen, test your predictions using a battery, piece of wire, and a flashlight bulb. To predict your answer write Y (yes) or N (no) on the blank beneath each picture.

1

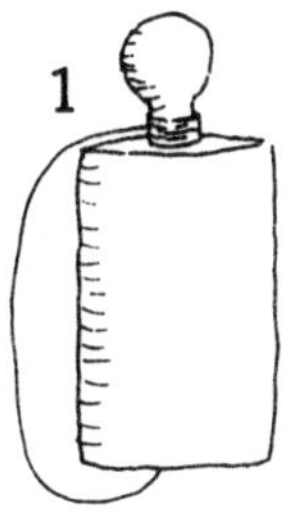

2

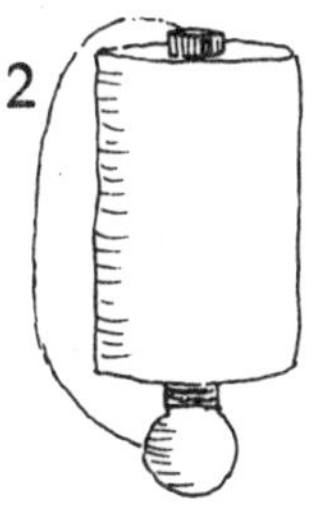

3

4

5

Using two pieces of wire, which bulbs will light?

6

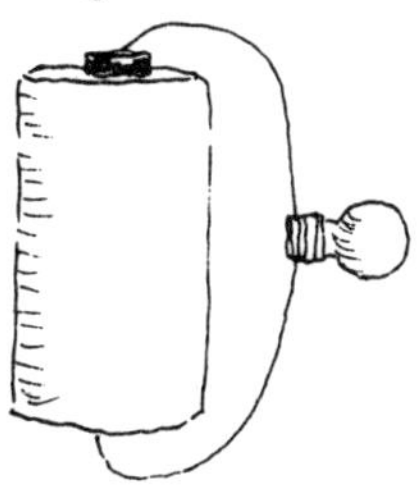

7

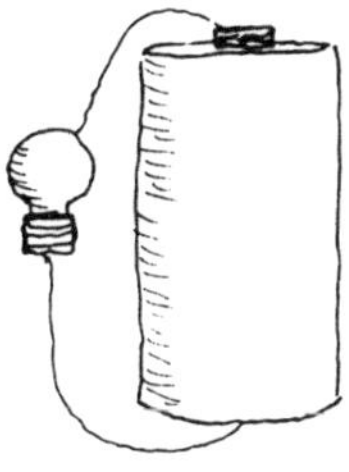

8

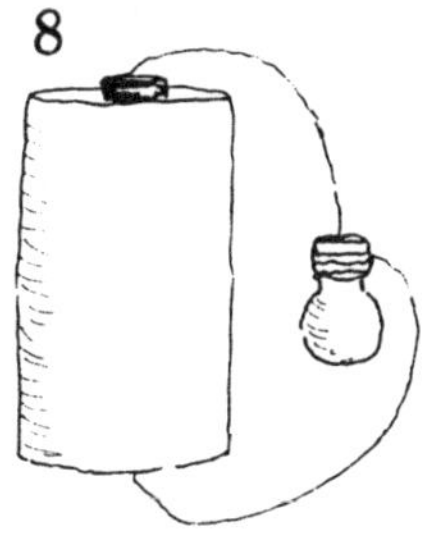

Name ____________________

CIRCUIT CARDS

Using a tester, determine which terminals on the electric circuit card are connected to form a complete circuit.

MATERIALS: 5" x 8" index cards
Aluminum foil
Hole puncher
Cellophane tape
Circuit tester

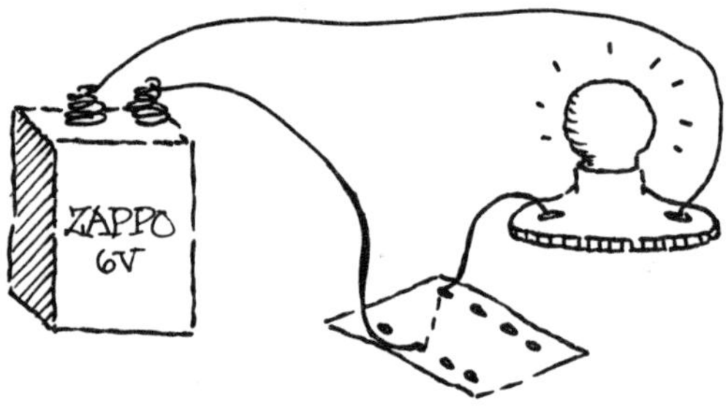

DIRECTIONS:

1. Punch four or five holes along each edge of an index card.

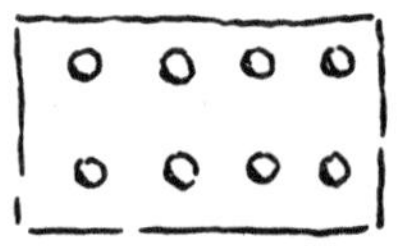

2. Cut strips of aluminum foil to connect some of the holes. Be certain all holes are covered with foil, even if they are not connected to another hole.

foil strip

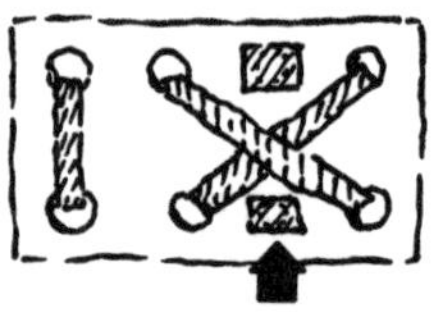

cover unconnected holes with squares of foil

3. Put tape over the foil strips so that no piece of aluminum foil overlaps or touches another piece of foil.
4. Tape another 5" x 8" card to the back to hide the foil arrangement. In the right-hand bottom corner of this card, show the pattern of foil connections used.

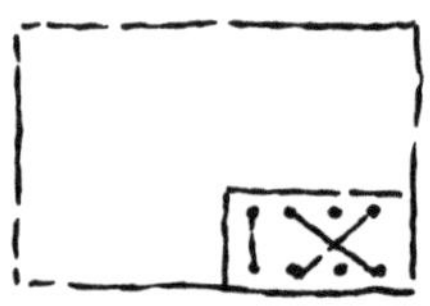

5. Exchange puzzles with other students. Using a tester, find the pattern of foil connections inside each card. Mark your answers on an answer sheet and then check the back of each card to see if you were correct.
6. For identification, write your name in the center of your card.

Name ______________________________

CIRCUIT CARD ANSWER SHEET

DIRECTIONS:

Draw lines to show how foil strips are connected on each card. Under each card, write the name on the card.

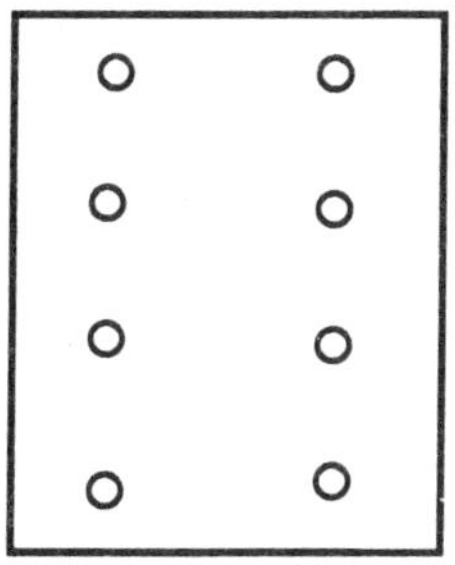
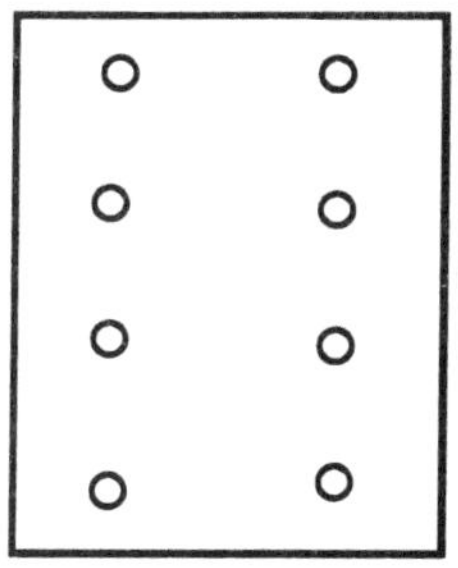
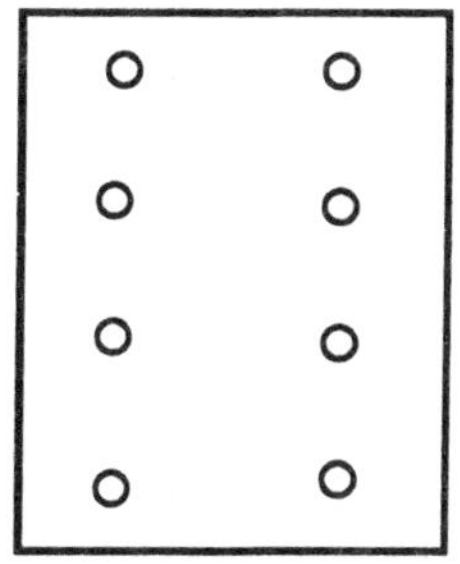
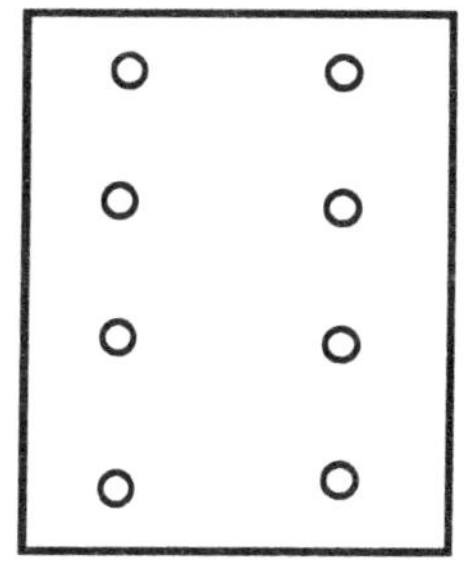

Name   Name   Name   Name

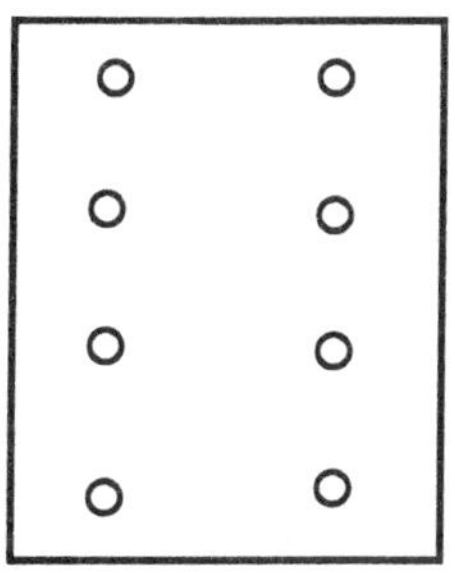
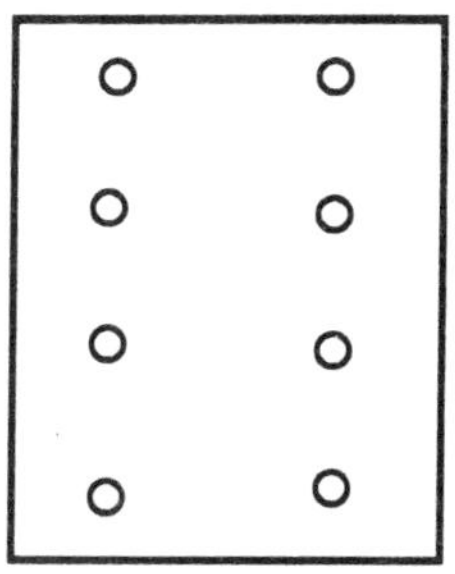
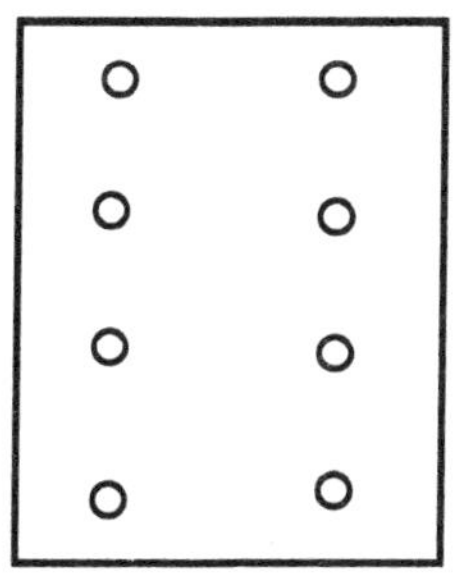
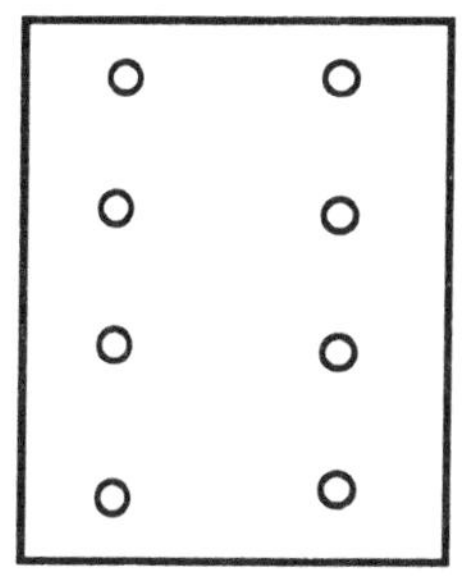

Name   Name   Name   Name

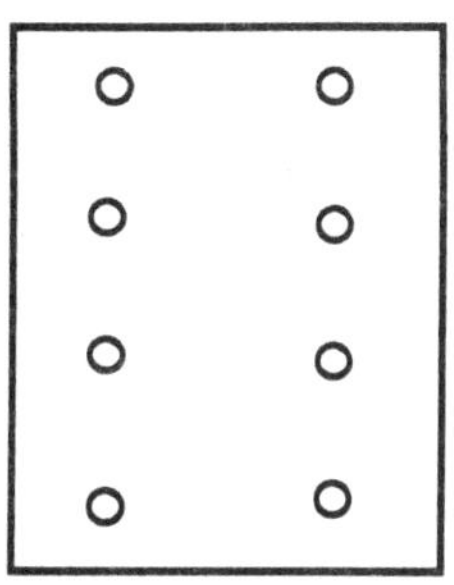
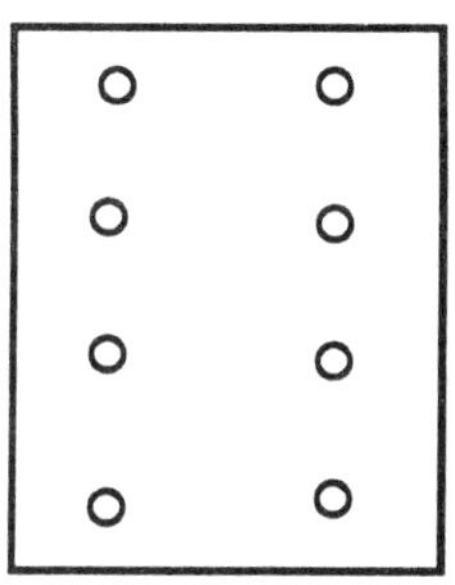
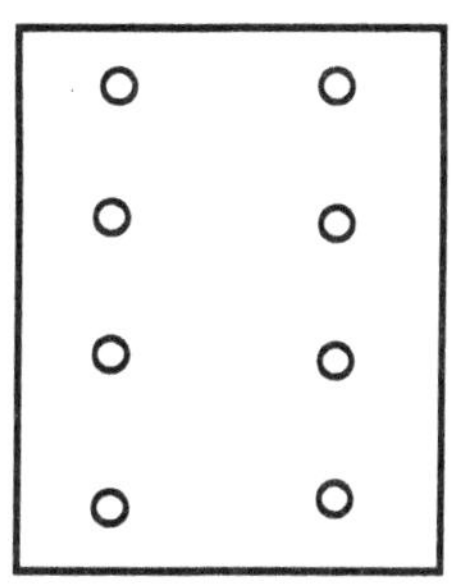
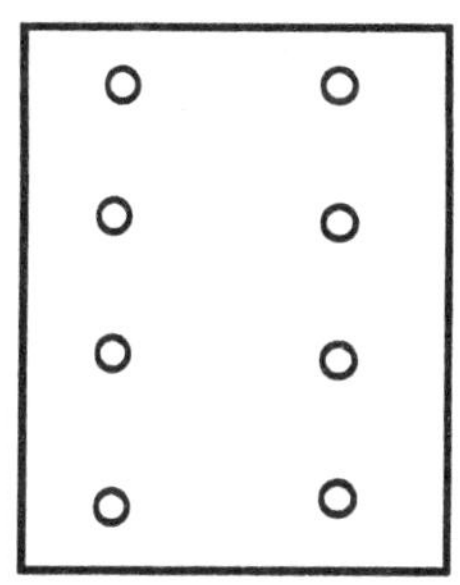

Name   Name   Name   Name

Name ______________________________

ELECTRIC LINGO

People who work with electricity use special symbols to represent batteries, wires, bulbs, and circuits. The symbol for a battery tells which way it is facing. When more than one battery is used in a circuit it is important to know which direction the battery is facing. Different things will happen depending on which way the battery faces. Square corners are used to make the drawing neat. They do not mean the wire has to have square corners.

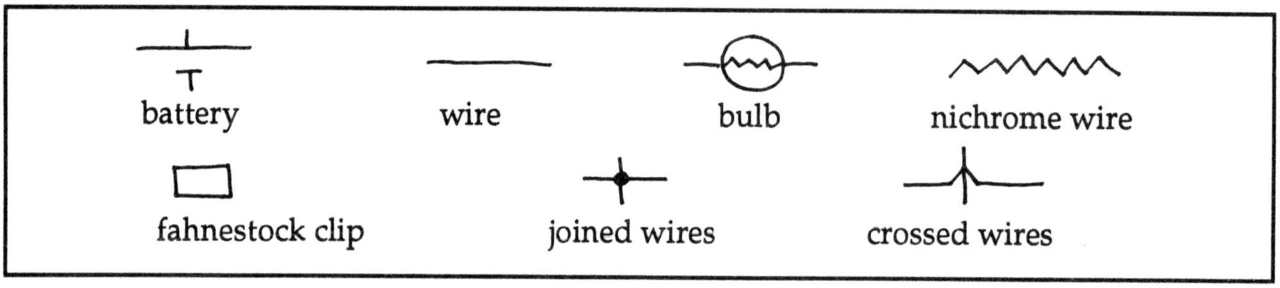

EXAMPLE:

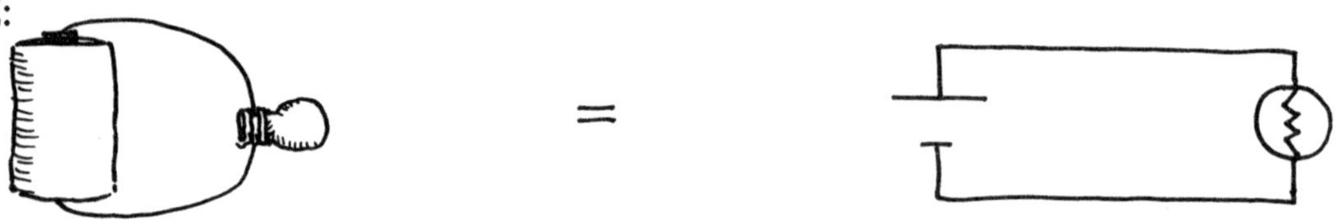

Draw the following circuits using the special symbols. Then try each of the circuits and tell whether the bulb will light.

1. 

2. 

3. 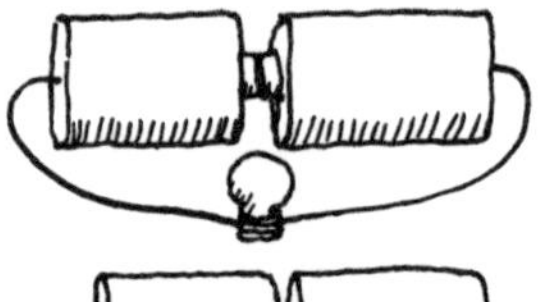

4. 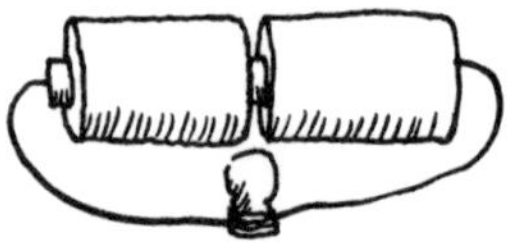

5. 

Name ______________________________

PARALLEL AND SERIES CIRCUITS

PARALLEL CIRCUIT

Add bulbs to a circuit as shown:

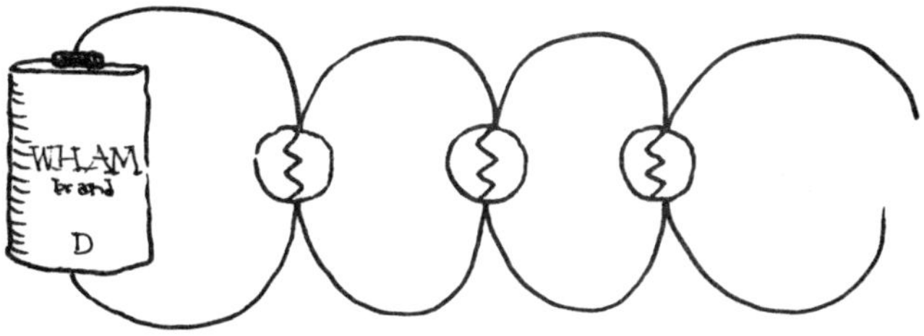

Draw this circuit using ELECTRIC LINGO

QUESTIONS:

1. Does the brightness of each light change as more bulbs are added?

   ____

2. What happens to the other bulbs when one bulb is removed? Why?

   ____

SERIES CIRCUIT

Add bulbs to a circuit as shown:

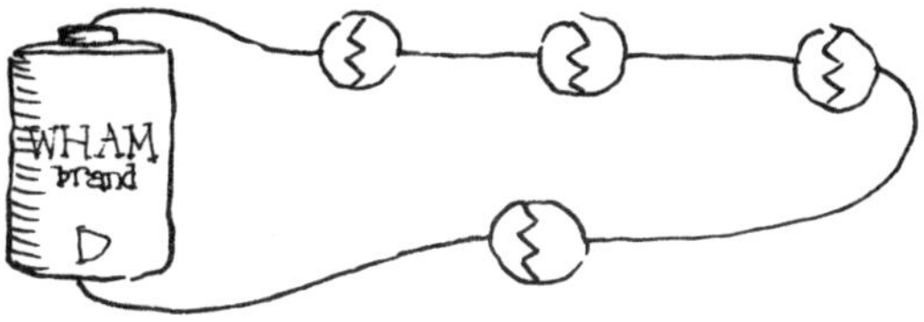

Draw this circuit in ELECTRIC LINGO

QUESTIONS:

1. Does the brightness of the lights change as more bulbs are added?

   ____

2. What happens to the other bulbs when one bulb is removed? Why?

   ____

3. What kind of a circuit would be best to use for Christmas tree lights? Why?

   ____

Name ______________________________

CONDUCTORS AND NONCONDUCTORS

Test different materials to see which ones will conduct electricity. Test each material by placing it across the nails or switch. If the material conducts electricity, the light bulb will light.

Place test materials here.

Try testing the following:

different metals	salt water	chalk
wood	rubber	fabric
plastic	glass	cardboard
cotton	leather	

CONDUCTORS	NONCONDUCTORS

QUESTION:

What similarities do you notice about the things that will conduct electricity?

____

____

Name ______________________________

ELECTROMAGNETS

Test the strength of your electromagnet by seeing how many paper clips it can pick up.

Some Things to Try:

- Experiment with nails made out of different metals. Does one kind of metal work better than another? ______________________________
- What happens if you connect the batteries in a series? ______________________________

  ____
- What happens if you connect the batteries in parallel? ______________________________

  ____
- Does it make any difference if the turns of wire are messy or neat? ______________________________

  EXPLAIN ______________________________

Make a list of all the materials you can find that are attracted to your electromagnet.

MATERIALS ATTRACTED TO AN ELECTROMAGNET	
1.	4.
2.	5.
3.	6.

QUESTIONS:

1. Are the materials that are attracted to an electromagnet the same as those attracted to a regular magnet? ______________________________
2. Name two things an electromagnet can do that a regular magnet cannot do.

   ____

   ____

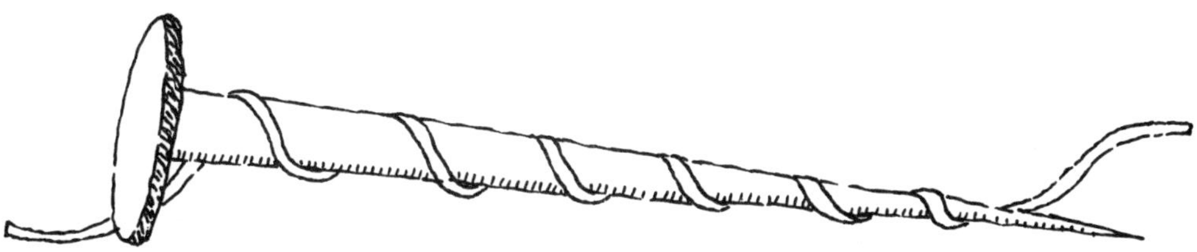

ELECTROMAGNETS

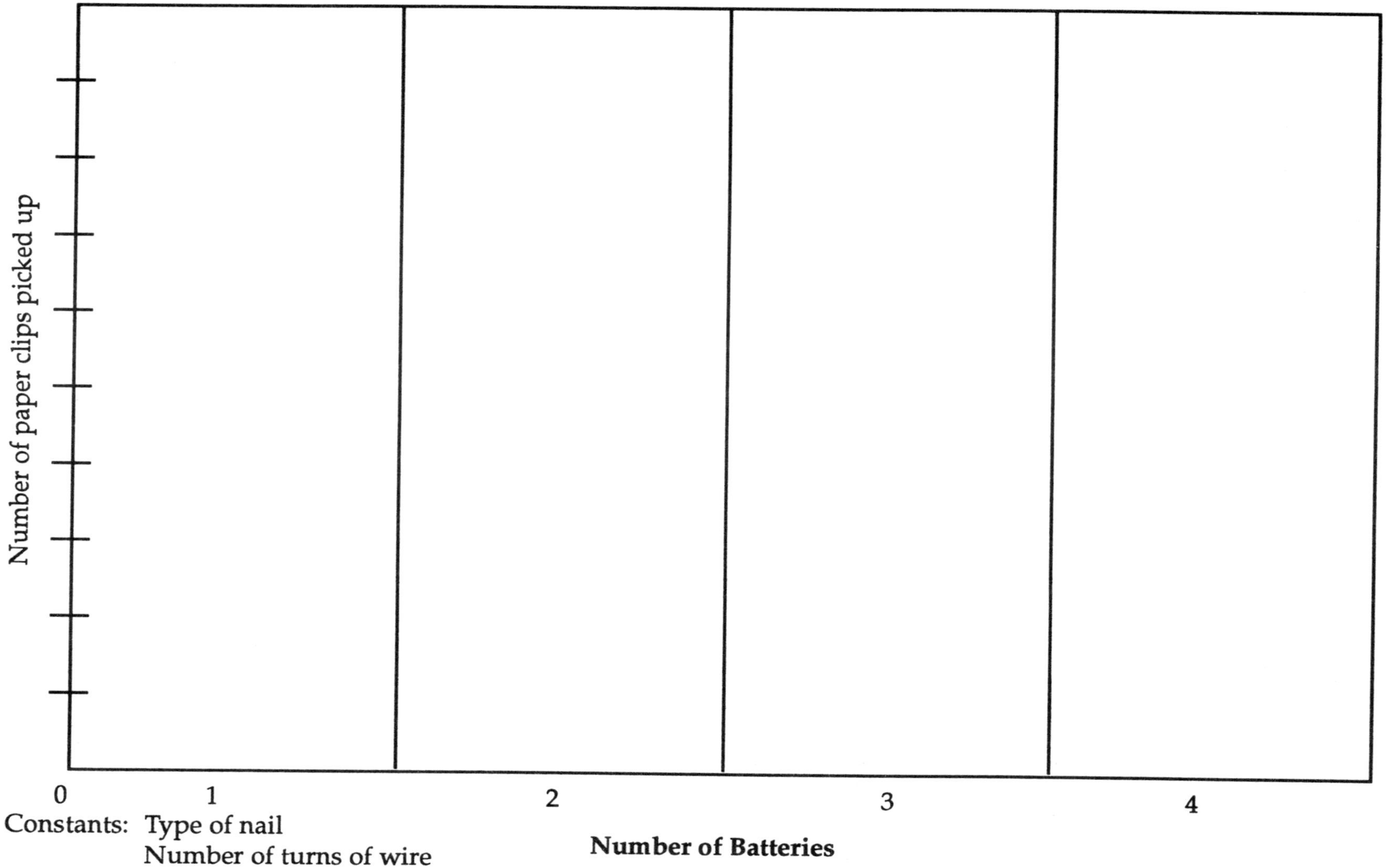

Constants: Type of nail
Number of turns of wire
Size of nail

Name ____________________

Name ______________________________

LET'S MAKE A BATTERY

Batteries make an electric current by using chemicals. The chemical causes electrons to collect on the zinc electrode inside the battery. Then the electrons move through the wire to the copper tube, creating a current.

MATERIALS: Glass jar with a wide top
Vinegar
Strip of zinc
Copper pipe
Two leads

DIRECTIONS:

1. Fill the jar with vinegar (acetic acid).
2. Stand the copper pipe and the strip of zinc in the vinegar.
3. Attach the end of one lead to the zinc.
4. Attach the end of the other lead to the copper.
5. Connect the open end of each lead to a small bulb (#48).

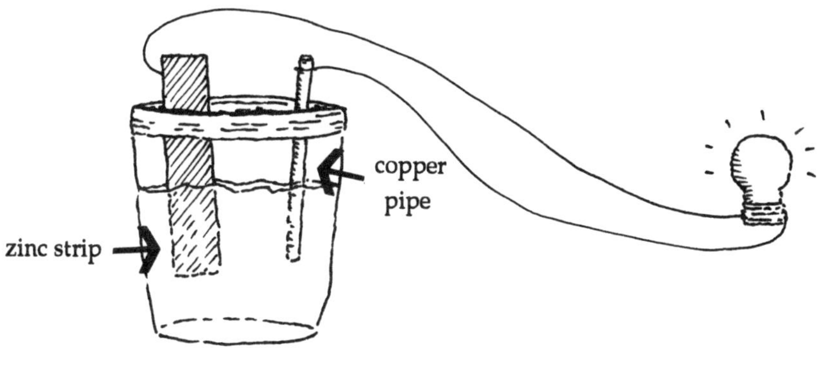

QUESTIONS:

1. What happens when the zinc and copper are taken out of the solution?

   ____

2. What happens when only the copper is taken out of the solution? Only the zinc?

   ____

Light

CHAPTER 2

How Light Travels
Page 22

TEACHER NOTES

Light is a form of energy that you can see. It is the only visible part of a wide range of electromagnetic radiation. Microwaves, X rays, radio waves, infrared radiation, ultraviolet light, and gamma rays are all forms of electromagnetic radiation, but they do not have the same wave length as light so they are not visible like light. The visible spectrum is only a portion of the electromagnetic spectrum.

Light sometimes behaves like a particle and sometimes acts as a wave. It travels in straight lines.

Laser light travels in one direction in a narrow beam. It travels in a straight line and does not spread out. The waves are all one color (same wavelength) and can concentrate the power and energy of the light.

Some lasers fire very short-lived pulses of light while others give off steady, long-lasting beams. Because of the power of laser beams, they have been used to perform surgery, especially on the eye. They can cut better than a knife and have been used to unplug blocked arteries, destroy tumors, and remove birthmarks.

Colors of Light
Pages 23 and 24

All of these colors (red, orange, blue, indigo, and violet) make up white light. Each color has its own special amount of energy and each color light has a different wavelength. Red has the longest wavelength and the lowest frequency of visible light. Violet has the shortest wavelength and the highest frequency of visible light.

RED	ORANGE	BLUE	INDIGO	VIOLET

Teacher Demonstrations

1. Observing the Colors of Light

Put a pan of water on a flat surface. Hold a mirror under the water at a slant. Darken the room and shine a bright flashlight on the mirror. Move the flashlight around until you see a reflection of light on the ceiling or wall. Hold up a piece of white paper to catch the colors.

2. Colored Soap Bubbles

Bend a length of thin flexible wire into any shape. The shape can be geometric, an animal, or free form.

Make a strong soap solution of dish washing liquid and water. Add two to three tablespoons of sugar to make the solution thicker. Fill one half of a bucket.

Dip the wire shapes into the solution. Slowly draw the wire out of the solution and examine the colors.

3. Observing the Spectrum

Make a one-half inch slit in a piece of paper or card and tape the card on a glass of water. Stand the card in front of a sunny window on a sheet of white paper. Sunlight passing through the slit will be refracted by the water. The spectrum will be produced on the paper.

DIRECTIONS:

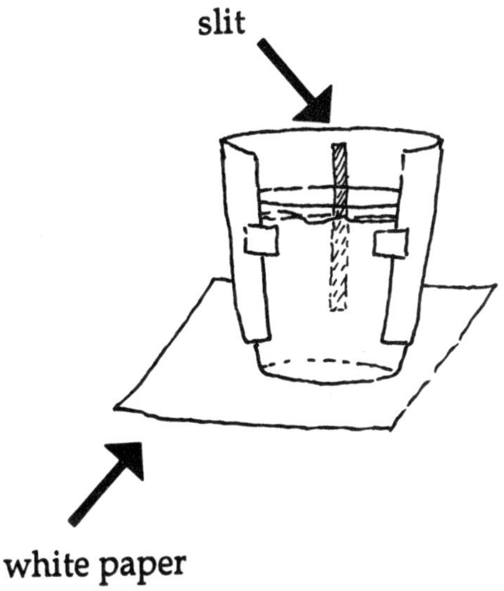

Use student activity sheet "Colors in Light" to experiment with mixing different colors and making white light. "Disk Patterns" can be used to put on cardboard disks to experiment with colors.

Reflections on Different Surfaces
Pages 25–27

Reflection is the bouncing of light off a surface. Different surfaces reflect light in differing degrees. A smooth, shiny, light-colored surface will reflect light the best. Dark-colored or rough-textured surfaces absorb most of the light. Mirrors, with their smooth, shiny surfaces, cause the light to bounce off in exactly the same pattern it arrives, reflecting an accurate picture of the image.

"Reflections on Different Surfaces" examines how different surfaces and different colors reflect light. Results from this activity sheet should be recorded on the two "Reflections" charts following it.

Observing Images in Mirrors
Pages 28 and 29

"Observing Images in Mirrors" provides additional activities to experiment with the reflection of light.

Solar Reflector Oven
Page 30

This activity sheet provides directions for making a solar reflector oven from everyday materials—an oatmeal box and aluminum foil. On a warm sunny day, the ovens may be used to soften marshmallows! Students observe how heat is captured by reflecting the sun's rays into the oven.

Refraction
Page 31

Refraction is the bending of light rays caused when light passes from one substance to another. The bending is caused because light travels at different speeds through different materials. Light travels at 186,000 miles per second. When it enters glass or water it slows down and changes direction.

The "Refraction" activity sheet, containing the bent pencil and the magic penny activities, provides students with some simple experiments to observe the bending of light as it passes through a liquid.

Can Liquids Magnify? Page 32

Liquids magnify objects because the light rays are refracted by the liquid the same way they are by a lens. Have students test the magnification of objects with three different liquids using the "Can Liquids Magnify?" activity sheet.*Lenses* are pieces of curved glass that reflect light in a precise way. They are used in cameras, microscopes, telescopes, glasses, and other items.

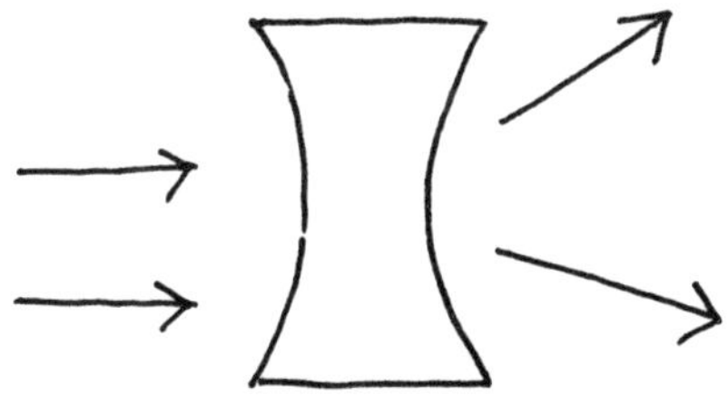

Concave lenses or *mirrors* are thinner in the middle than on the ends (dished) and spread the light rays as they pass through. Concave lenses make objects appear smaller.

Convex lenses or *mirrors* are thicker in the middle (bowed) and converge the rays of light passing through. The rays cross at a point called the *focus*. Convex lenses make objects look larger. A magnifying glass is a simple convex lens.

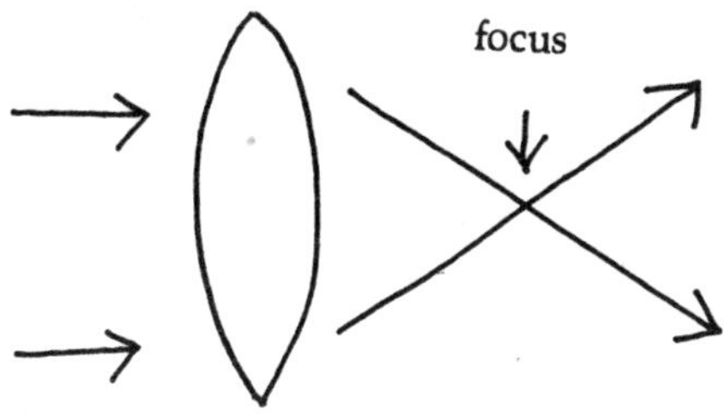

Transparent, Translucent, or Opaque? Page 33

Material that allows light to pass through is *transparent*. A *translucent* material allows light to pass through but images cannot be seen clearly. An *opaque* material does not allow any light to pass through.

Using this activity sheet, students will experiment with a variety of materials and classify them into one of the categories above.

Shadows Pages 34–37

Students will experiment with the length of shadows under various circumstances. The three charts can be used to record information when experimenting with:

- Shadow length related to straw length.
- Shadow length related to distance between straw and bulb.
- Shadow length related to brightness of light.

Name ____________________

HOW LIGHT TRAVELS

MATERIALS: 4 index cards
Hole punch
Masking tape
Battery and flashlight bulb
Paper cups

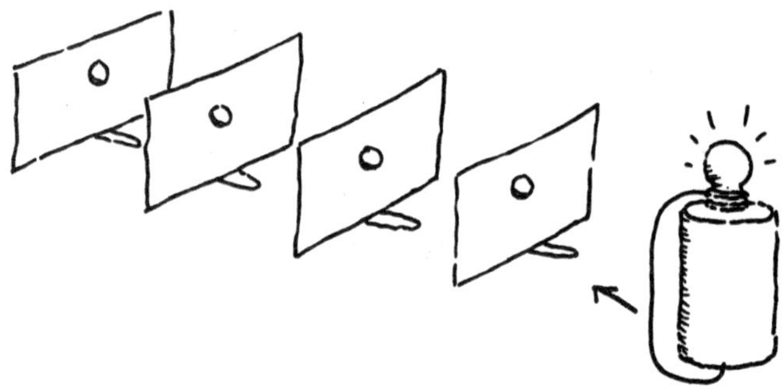

DIRECTIONS:

1. Punch a small hole in the center of each index card. Use paper clips to hold the cards. Attach the paper clips with masking tape and line the cards up in a row.
2. Put the light source three inches beyond the last card.
3. With one eye, look through the hole in the first card. Move the cards so you can see the light.

QUESTIONS:

1. What kind of a path do the holes make?

____

____

2. Can the cards be lined up any other way so you can see the light?

____

____

3. Must the holes always be in a straight line? (Explain)

____

____

Try using a lens (magnifying glass) to bend the light and see what happens. Explain what happened.

____

____

____

____

Name ______________________

COLORS OF LIGHT

What colors can you make on a spinning disk if you use only red, green, and blue?

MATERIALS: White cardboard circles
Sharpened pencil
Red, blue, and green markers

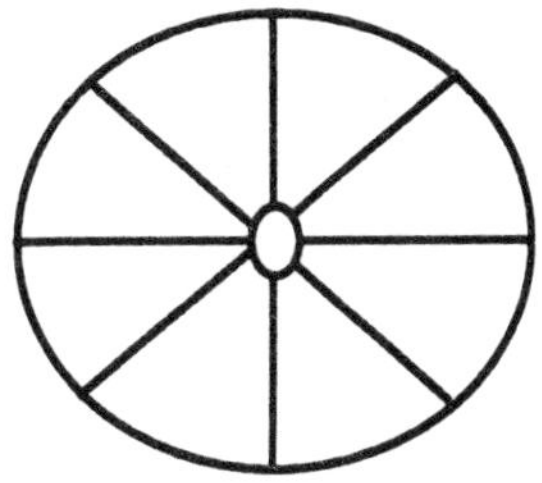

DIRECTIONS:

1. Cut out cardboard disks using the size of the patterns on the next page.
2. Make a hole to fit the pencil in the center of each disk.
3. Color different parts of each disk red, green, or blue.
4. Spin the disk on the pointed end of a pencil.

WHAT COLORS DO YOU SEE?

Color		Recipe
________	=	______________________
________	=	______________________
________	=	______________________

TRY THESE COLOR COMBINATIONS. WHAT COLORS DO YOU SEE?

½ red + ½ green = ______________________

½ red + ½ blue = ______________________

½ blue + ½ green = ______________________

MAKE SOME NEW COLORS.

(You can leave some areas white but use only red, blue, and green.)

GRAY = ⅓ red + ⅓ blue + ⅓ green
ORANGE =
YELLOW =

Color		Recipe
________	=	______________________
________	=	______________________

*MAKE WHITE LIGHT! Divide the disk into seven parts and color each section one of the colors in the spectrum. Spin the disk!

DISK PATTERNS

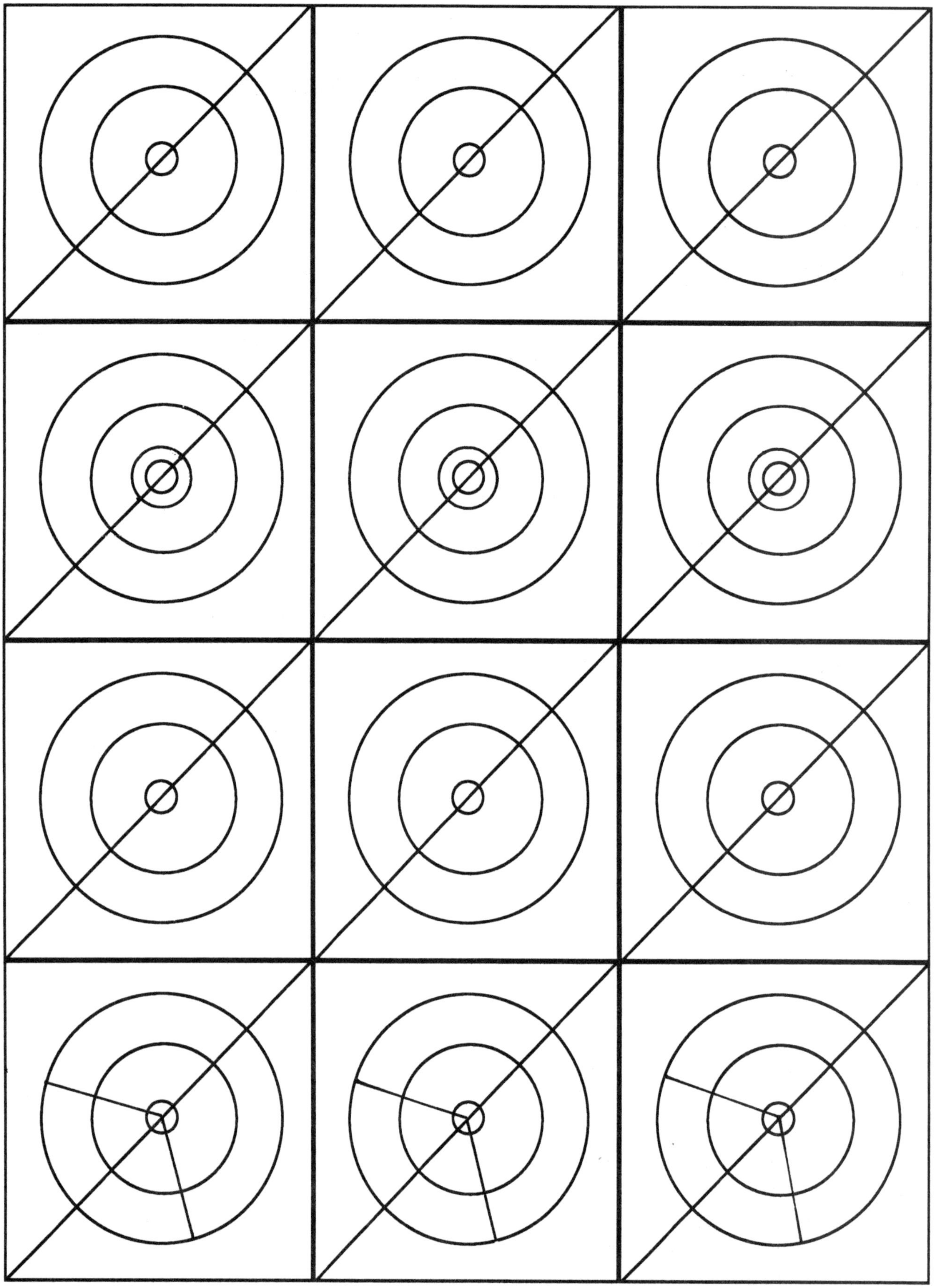

Name ______________________

REFLECTIONS ON DIFFERENT SURFACES

Different surfaces reflect light in different degrees. Different colors reflect light in different degrees as well.

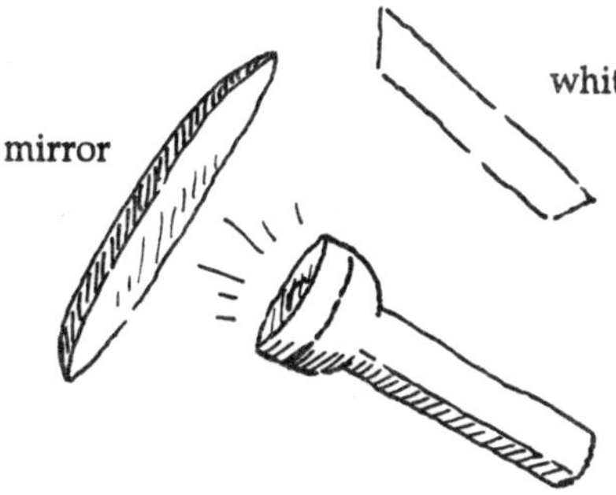

MATERIALS: Flashlight
Flat mirror
White construction paper
Black construction paper

DIRECTIONS:

1. Darken room.
2. Place mirror and card or two cards at an angle to each other.
3. Shine the flashlight onto the mirror and see what the reflection (on the white card) is like.
4. Write down what you see reflected on the white card.

   A. Illustration above with a mirror.
   B. Replace the mirror with a black card.
   C. Replace the mirror with a white card.

Results: (Describe reflection on white card.)

A. ______________________

B. ______________________

C. ______________________

QUESTIONS:

1. Which surface reflected the light onto the white card the best?

   ____

2. Which surface reflected the light the poorest? Why?

   ____

3. Based on this activity, which surface and color do you think reflects light the best?

   ____

More Activities to Try:

Will the light reflect differently with other colors? Try other colored paper in place of the mirror. How will the light be reflected from different surfaces? Try smooth, pleated, and wrinkled aluminum foil instead of the mirror.

Record your observations for these experiments on the "Reflections" chart pages.

Name ______________________

Reflections

Colors	Poor Reflection	Good Reflection	Strong Reflection
Red			
Green			
Blue			
Yellow			
Purple			
Brown			

How will different colors reflect light? ______________________

Name ____________________

Reflections

Surface	Poor Reflection	Good Reflection	Strong Reflection
smooth foil			
wrinkled foil			
pleated foil			

How will different surfaces reflect light? ____________________

Name ______________________

OBSERVING IMAGES IN MIRRORS

MATERIALS: Flat mirror
2 file cards
Pencil
Shiny metal spoon

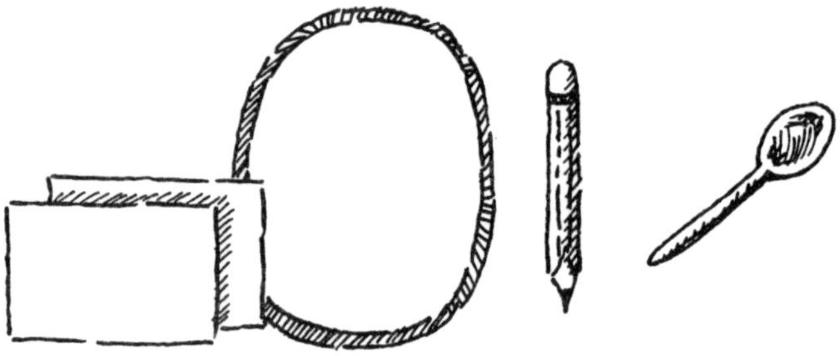

DIRECTIONS:

1. Write the capital letter *F* on a file card and hold the card in front of a flat mirror. Observe the image of the letter and draw it in box 1 below.
2. Write a backward capital letter *F* on a card. Hold the card in front of the flat mirror. Observe the image and draw it in box 2 below.
3. Print your name on a file card. Hold the card in front of the mirror. In box 3 write what you see in the mirror.
4. Look at the capital letter *F* in the back of the spoon. Write what you see in box 4 below.
5. Look at the letter *F* in the front of the spoon. In box 5, copy the image you see.
6. Look at your image in the front and back of the spoon. Describe what you see. Write your answer in #6 below.

IMAGES IN A FLAT MIRROR

1. [box] Letter F
2. [box] Backward F
3. [box] Printed Name
4. [box] Image of F in back of spoon
5. [box] Image of F in front of spoon
6. **YOUR IMAGE**
Back of spoon ______________
____
Front of spoon ______________
____

Name ______________________________

OBSERVING IMAGES IN MIRRORS

QUESTIONS:

1. What happens to the image of an object when it is placed in front of a flat mirror?

   ____

   ____

2. What is the difference between the image in the flat mirror and the image in the curved mirror?

   ____

   ____

Some More Activities to Try:

- What will happen to the image in a curved mirror if you put a flat mirror in front of it? (Put a flat mirror in front of the image on the spoon.) First predict what will happen. Try it and see. Record your answers below:

  Predictions: ______________________________

  ____

  What happened: ______________________________

  ____

- Write the words OX and EAR on a piece of paper. Place the paper in front of the spoon. Tell what happens to each image.

  OX ______________________________

  EAR ______________________________

  (Why doesn't OX appear upside down?)

Name ______________________________

SOLAR REFLECTOR OVEN

MATERIALS: 1 round oatmeal box, 1 shoe box, Marshmallows, Aluminum foil, Coat hanger, A WARM, SUNNY day, Chocolate bar, Graham crackers

DIRECTIONS:

1. Cut the front out of the shoe box to make a frame. Poke a hole in each top front corner.
2. Mark a point halfway along the radius on both ends of the oatmeal box and punch a hole there.
3. Cut the oatmeal box in half down its length.
4. Line the box with aluminum foil.
5. Unbend the coat hanger and push it through the holes on one side of the shoe box frame and the oatmeal box.
6. Put two or three marshmallows on the other end of the coat hanger and push that end in the other side of the oatmeal box and the shoe box frame.
7. Place cooker in the sun.

Describe what happens: Explain how the reflection of light makes this oven work.

____

____

____

____

Put your marshmallows and a piece of chocolate bar between two graham crackers. ENJOY!

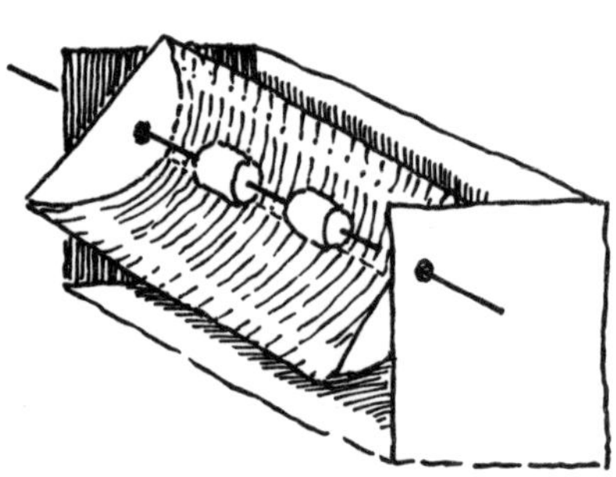

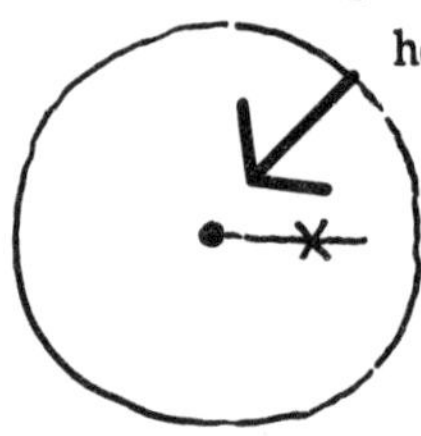

Name ______________________________

REFRACTION

Refraction is the bending of light that occurs when light passes from one transparent substance to another.

THE BENT PENCIL

DIRECTIONS:

1. Put a pencil into a glass of water.
2. Look at the pencil.
3. Draw the pencil in the glass of water as it appears to you.
4. Describe how the pencil looks. Explain why.

Write your answer here:

____

____

____

MAGIC PENNY

MATERIALS: 1 empty clear plastic cup
1 cup water
1 penny

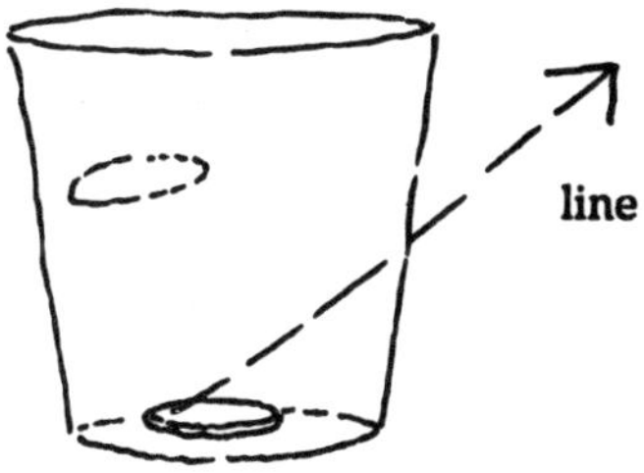

DIRECTIONS:

1. Work with a partner.
2. Put the penny into the empty cup.
3. Move your head down until you no longer see the coin.
4. Keep your head still while your partner slowly fills the cup with water.

QUESTION:

Describe what happens to the penny. Explain why.

Write your answer here:

____

____

____

Name ______________________________

CAN LIQUIDS MAGNIFY?

MATERIALS: 3 letter *B*s all the same size (approx. ⅛-¼") each written on a small piece of paper
2 medicine droppers
Water
Cooking oil
Piece of cardboard
Piece of white paper
Piece of plastic wrap

DIRECTIONS:

1. Glue a piece of white paper to the cardboard.
2. Glue the three letters to the paper.
3. Cover the board and letters with plastic wrap. Make sure the plastic wrap is flat and tight.
4. Put a drop of water on one *B*.
   Put a drop of oil on another *B*.
   Put nothing on the third *B*.
5. Measure the *B* with nothing on it.
   Measure the other *B*s.

Record the results below.
Explain how a liquid can change the size of the letter.

____

____

____

	Same Size	Larger	Smaller
water			
oil			

(Can you think of some other liquids to try?)

Name ____________________

TRANSPARENT, TRANSLUCENT, OR OPAQUE?

Transparent – material through which light is allowed to pass.

Translucent – material that allows light to pass through but scatters the light so things cannot be seen clearly through the material.

Opaque – material through which light is not allowed to pass.

Consider the following materials. Tell whether they would be transparent, translucent, or opaque.

house plant	frosted glass	cardboard
water	paper	air
stone	wood	wax paper
foil	plastic wrap	cheesecloth
cotton cloth	leather	plastic

DIRECTIONS:

Working with a partner, test each of these materials by trying to look through them to someone on the other side. Write each item in the appropriate column.

Transparent	Translucent	Opaque

(Can you think of some more materials to test?)

Name ___________________________

SHADOWS

MATERIALS: Source of light
Large sheet of white paper
Straw
Cellophane tape

DIRECTIONS:

Tape the straw to the sheet of paper. Place the bulb in a position so the straw makes a long shadow.

QUESTIONS:

1. What happens to the length of the shadow if you halve the length of the straw?
____
2. What happens to the length of the shadow if you quarter the length of the straw?
____
3. Can you figure out a way to calculate shadow length from straw length? (Use the "Shadows" graph to record your observations.)
____
4. Does the brightness of the light affect the shadow?
____
5. Is the length of the shadow affected by the distance between the straw and the bulb?
____

Explain your experiment in writing and record your findings on the "Shadows" graphs.

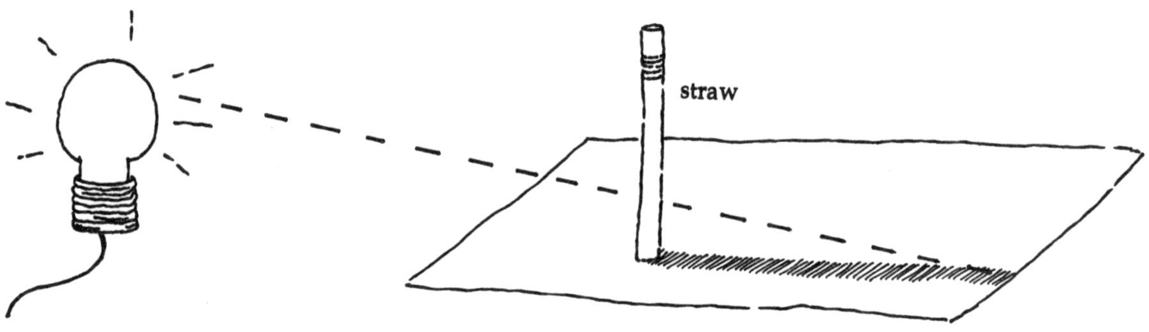

Name ____________________________

SHADOWS

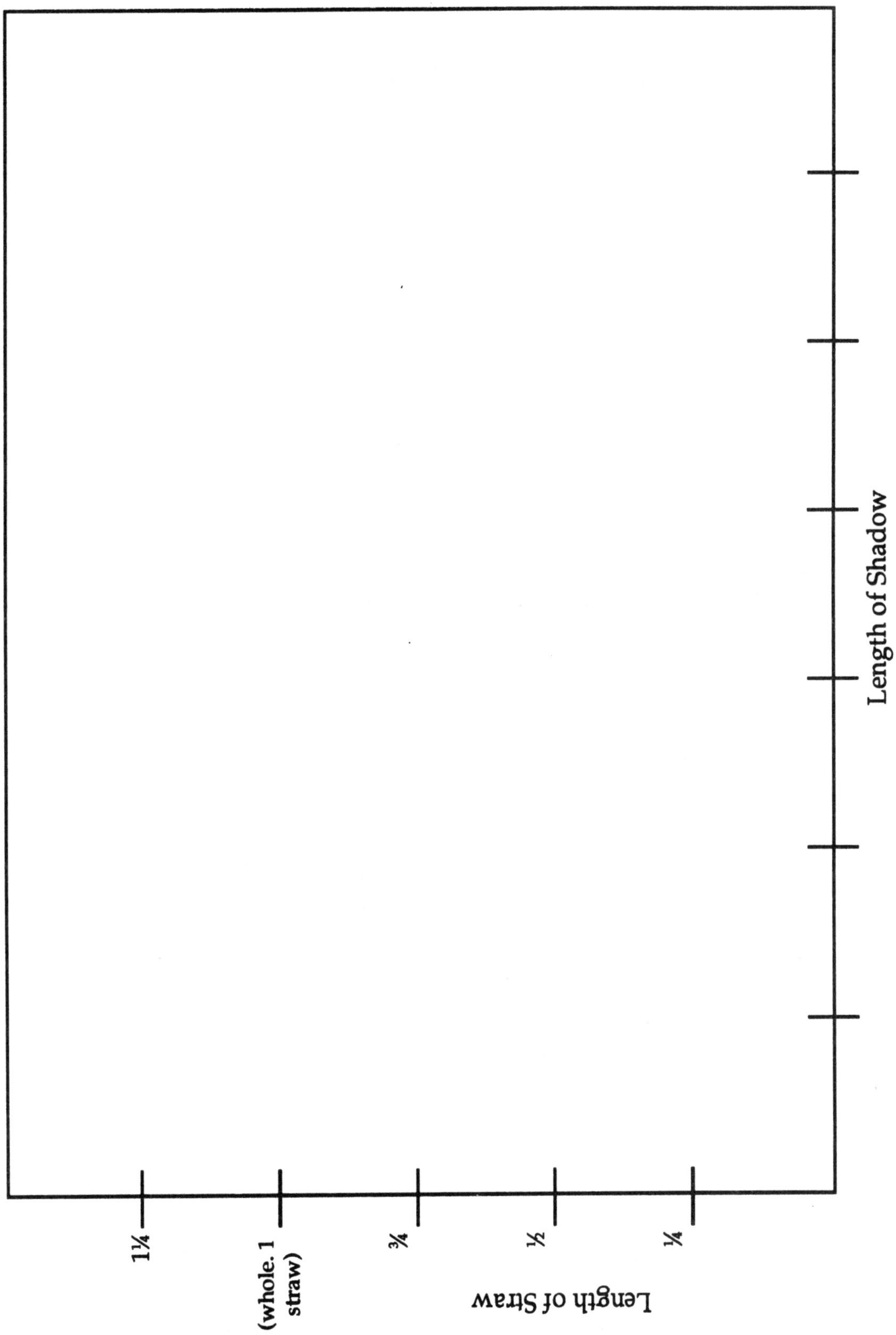

Calculating shadow length from straw length ____________________________

____

Name

SHADOWS

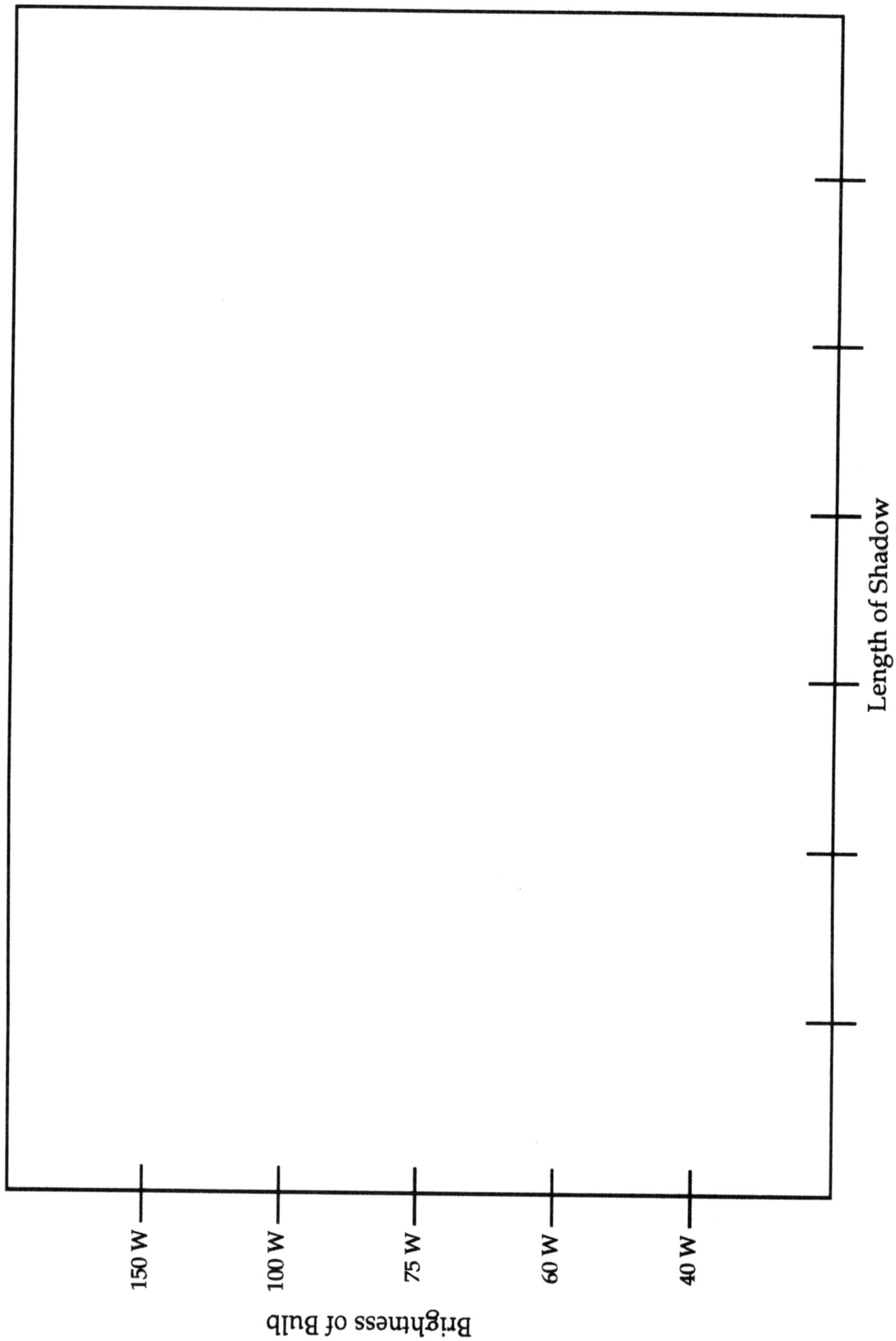

Does brightness of light affect the shadow length?

Name ____________________________

SHADOWS

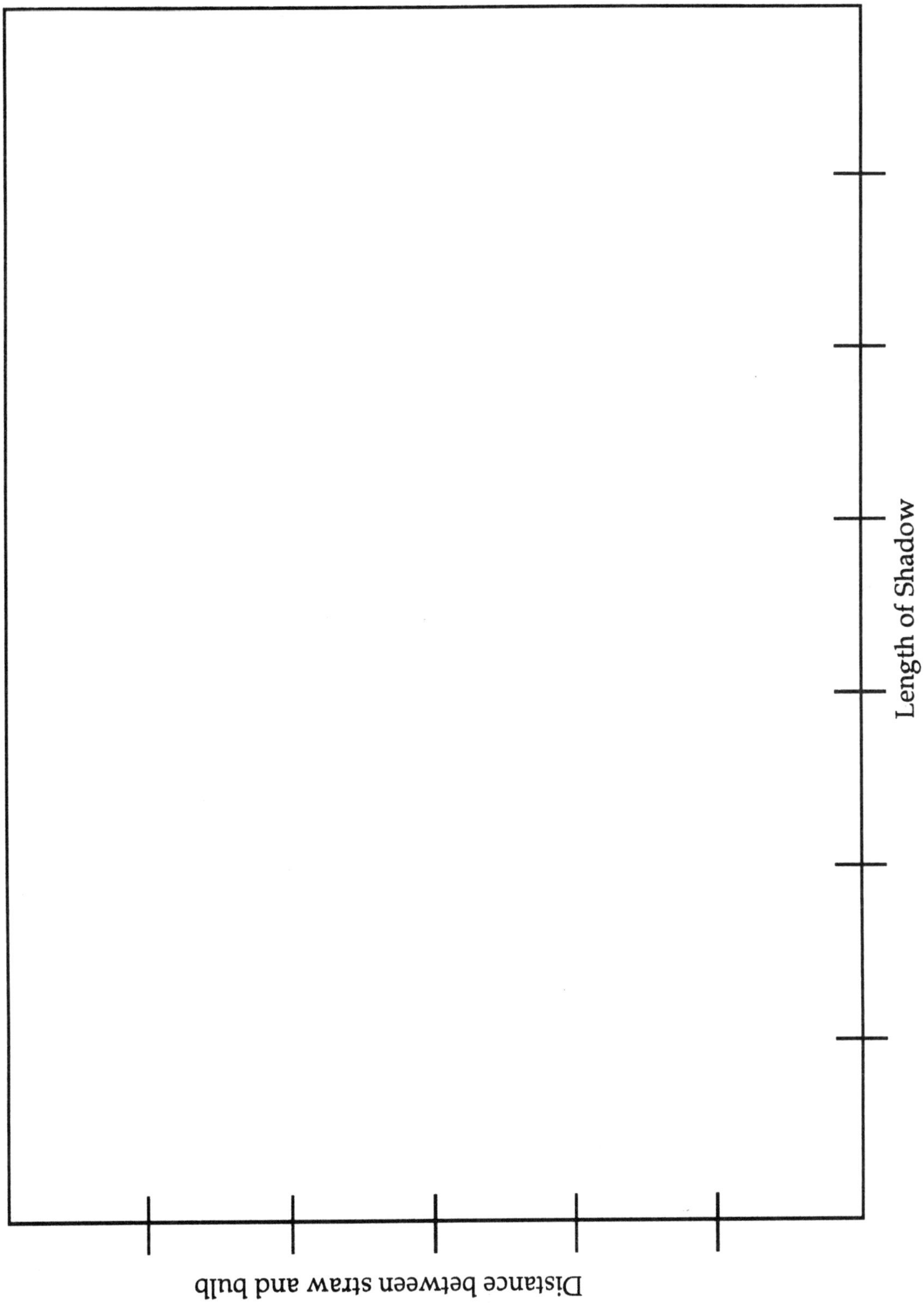

Does distance between straw and bulb affect length of shadow? ____________________________

____

Sound

CHAPTER 3

TEACHER NOTES

HOW SOUND TRAVELS

Sound waves move by alternately squeezing or stretching. They do not go up and down like light waves. Sound waves move like a spring. Sound waves travel more slowly than light waves. They cannot travel through space or through a vacuum. They must travel through matter because matter can vibrate.

DEMONSTRATIONS THAT ILLUSTRATE HOW SOUND TRAVELS.

1. Use a Slinky to visualize the movement of a sound wave.
2. Stretch a balloon over the end of an empty soup can from which the ends have been removed. Cut the open end of the balloon to enlarge the opening. Secure the balloon with a rubber band, stretching the balloon tightly across the opening. Place grains of rice on the balloon and hold the can over a radio on which the volume has been turned up high. The rice will bounce and vibrate from the sound waves that are travelling through the air.

3. Place a 12-inch ruler over the edge of a desk. Let 9 inches extend over the edge. Snap the extended end. What do you see and hear? (A low humming sound as the ruler vibrates.) Now let only 6 inches extend and snap the ruler again. How is what you see and hear different from the first time? (Ruler vibrates faster and the pitch is higher than the first time.)

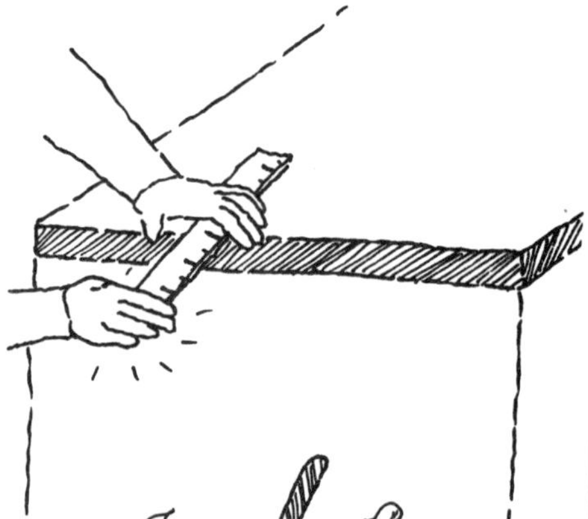

4. Hit a tuning fork with a hammer and place it in a glass of water. What happens to the water and why? (The water vibrates so fast that some of it splashes out of the glass. It shows that sound makes things vibrate. The water is being splashed out in all directions because sound waves travel out from a vibrating body in all directions.)

Jingling Jar
Page 41

Using the "Jingling Jar" activity, students will observe that the sound of the bell cannot be heard as clearly when hot water is added to the jar because there are fewer molecules in the warmer air. Sound travels best through substances where the molecules are close together.

Let students experiment to find out what happens when cold ice water is added to the jar. Why does the bell sound louder?

Sound waves travel well through air but they travel even faster through liquids and solids. Sound waves travel about 1,130 ft. per second through air and about 1,500 ft. per second in pure water. In solid steel they travel about 20,000 ft. per second. The sound travels fastest through solids because the molecules in a solid are closer together. Sound will not travel through a vacuum because there is nothing to vibrate. High and low pitched sounds travel at the same speed.

Sound also travels faster in warmer temperatures. An increase of one degree in temperature will cause an increase of two feet per second in travel rate.

ULTRASOUND–sound pitched higher than the human ear can hear.

INFRASOUND–sound pitched lower than the human ear can hear.

Some interesting facts!

1. Bats and dolphins can hear ultrasound more than six times the highest pitch that humans can hear.
2. Sonar on ships uses ultrasound echoes to determine the location of objects underwater.
3. In industry, ultrasound echoes can detect flows inside solid metal.
4. In medicine, sound waves can be used to produce a live image of the inside of your body. Ultrasound has no harmful side effects like X rays.
5. Ultrasound guns are used to break up kidney stones without surgery. The sound shakes the kidney stones to pieces. These guns are called *lithotripters*.

FREQUENCY–number of sound waves per second. Frequency is measured in *hertz* (Hz). The frequency of sound waves determines the pitch. Humans can hear down to 20 Hz and up to 20k Hz. Sound higher than 20k Hz is ultrasound.

High sound – high frequency
Low sound – low frequency

TONE–Different materials and different musical instruments produce different wave patterns so even though a violin and a flute play the same note, they sound different.

LOUDNESS–Depends on the energy in a wave. Loudness is measured in *bels* after Alexander Graham Bell. A decibel is equal to .1 bel.

Bottle Band
Shoe Box Guitar
Pages 42 and 43

Using the "Bottle Band" activity sheet, students can see the relationship between the amount of air that is vibrating and the pitch. The highest sound is made by the bottle with the least amount of air.

Students will discover with the "Shoe Box Guitar" activity that something vibrates faster when it is pulled tightly, causing a higher pitch.

Echoes
Page 44

ECHOES–Sound waves are reflected from hard, smooth surfaces. They are absorbed by soft, rough, or porous material. When sound waves hit a soft, textured material most of the waves are absorbed. Only a few are reflected.

SONAR—A sound echo. Ships use echoes to determine the depth of the ocean. Echoes underwater are also used to detect the location of anything from icebergs to enemy submarines.

Use the "Echoes" activity sheet to experiment with echoes while walking and to see how sound can be absorbed by different materials. Discuss the use of acoustical materials—curtains, acoustical tiles on ceilings, and material on walks.

Research ways acoustics in a room can be improved by the way a room is shaped or by the materials used in construction.

Name ______________________________

JINGLING JAR

MATERIALS: 8-oz. glass jar with tight lid
Hot water
Small bell
Tape

DIRECTIONS:

1. Suspend the bell in the jar by taping the string to the inside of the lid. Tie the other end of the string to the bell.
2. Shake the jar, listening to the sound of the bell.
3. Remove the lid of the jar (with the bell attached) and pour about 1" of hot water into the jar. Let it stand for a few seconds and put the lid back on.
4. Shake the jar again and listen to the bell.

QUESTION:

What happened to the sound of the bell when hot water was added to the jar? Explain why. HINT: What happens to the molecules in the air when the air gets heated?

____

____

____

____

Try This!

Add some cold ice water to the jar after it cools down. Does cold air affect the sound of the bell?

Name ____________________

BOTTLE BAND
(pitch)

DIRECTIONS:

1. Fill each of five empty soda bottles with increasing amounts of water. You can add food coloring to make each bottle of water a different color.
2. Blow across the top of each bottle.

QUESTIONS:

1. Which bottle makes the lowest sound?

 ____

 ____

2. Which bottle makes the highest sound?

 ____

 ____

3. What conclusions can you make about the amount of air in the bottle in relation to the sound?

 ____

 ____

Try different amounts of water in eight bottles until you can make a simple scale when you blow into consecutive bottles. Can you play a simple tune by blowing on the bottles? Number the bottles 1-8 and write down your tune.

MY TUNE

_____ _____ _____ _____ _____ _____

_____ _____ _____ _____ _____ _____

Name ______________________________

SHOE BOX GUITAR

MATERIALS: Shoe box with a lid
Toothpick
Pencil
50-cm. piece of fish line
Two 10-cm. pieces of wooden dowel

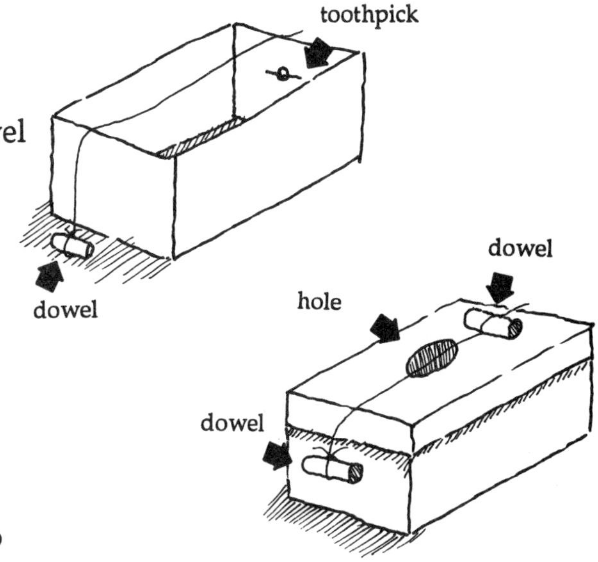

DIRECTIONS:

1. Remove the top of the shoebox and make a hole in one end of the box with your pencil. Put the fish line through the hole and secure it with a toothpick inside the box.
2. Put the lid back on the box and stretch the fish line across the top of the box. Put one piece of the dowel under the line near the top of the box. Tie the other dowel onto the other end of the line and let it hang over the edge of the box.
3. As you pull down on the dowel, pluck the string. Continue making the string tighter.

QUESTIONS:

1. What happens as you make the string tighter?

____

____

2. Cut a small hole in the lid under the string. What happens then?

____

____

3. What conclusion can you draw about the relationship of the speed of the vibration to the pitch (high-low)?

____

____

Name ______________________________

ECHOES

ACTIVITY I

DIRECTIONS:

Use leather shoes or shoes with taps. Have someone blindfold you and lead you to the other end of a hall at school. As you walk down the hall, listen to the sound of your feet. (Leave several doors open to the hall.)

QUESTIONS:

1. Do your steps sound different as you get near the end of the hall?

   ____

2. Where did you first hear an echo?

   ____

3. Could you tell when you passed open doors?

   ____

ACTIVITY II

MATERIALS: Rubber ball
Soft pillow
Large, clean metal bucket
Large piece of soft cloth

DIRECTIONS:

1. Throw the ball against a hard wall. What happens?
2. Throw the ball against a pillow. What happens?
3. Talk into the bucket. How does it sound?
4. Put a cloth around the inside of the bucket. Talk again. What happens?

QUESTION:

Why are musicians concerned with acoustics in practice booths? Why are curtains used in auditoriums?

____

____

CHAPTER 4 — Simple Machines

TEACHER NOTES

All machines are made up of various combinations of the six simple machines. The six simple machines are the lever, wheel and axle, pulley, inclined plane, screw, and the wedge.

Book Lift
Friction Fighters
Pages 49 and 50

Inclined Planes

An *inclined plane* is a flat, slanting surface used to raise heavy things without having to lift them vertically. A load can be moved up an inclined plane with less force than would be indicated by the weight of the object.

Friction is caused whenever two objects are rubbed together. Friction causes heat. The use of a lubricant can reduce the amount of friction. Starting friction is greater than sliding friction because more force is needed to start an object than to keep it sliding. Rolling friction is less than sliding friction and is the reason wheels and bearings are used in complex machinery.

Using the "Book Lift" activity sheet students can see how work becomes easier when using an inclined plane. "Friction Fighters" will allow students to experiment with different substances to see which lubricants best reduce the effects of friction.

Levers
Pages 51–56

A lever is used to decrease the amount of force required to lift a load by moving the fulcrum closer to the load. Although there is a decrease in force required, speed and distance are sacrificed. As force is decreased, the effort arm travels a greater distance and the load travels less distance. The farther away from a fulcrum (a fixed point) the force is applied, the greater its effect. There are three basic types of levers. The first-class lever has the fulcrum between the effort and the load. The second-class lever has the load between the effort and the fulcrum. In a third-class lever, the effort is between the load and the fulcrum.

LEVERS

F = Fulcrum (Effort shown on spring scale balance)

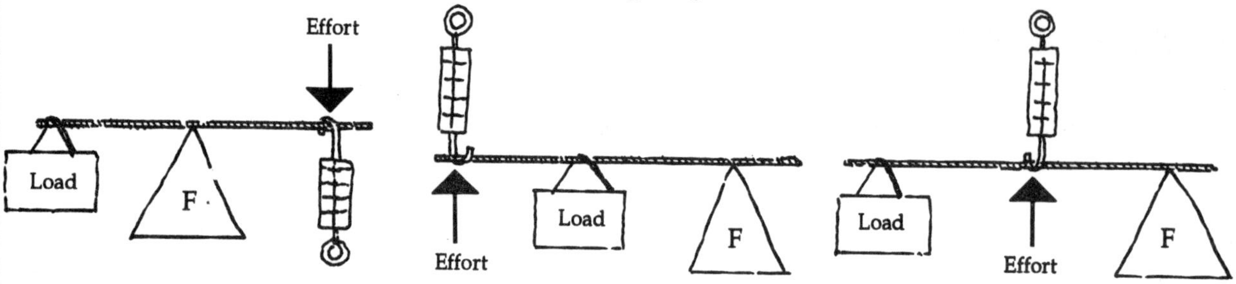

First-Class Lever	Second-Class Lever	Third-Class Lever
Pliers	Screwdriver	Tweezers
Scissors	Nutcrackers	Sugar tongs
Crowbars	Wheelbarrows	The lower arm

The following activity sheets provide activities for students to practice with first, second, and third-class levers.

"Lifting Loads with First-Class Lever"

"First-Class Lever (Force Measurements)"

"Relationship of Force and Distance Using First-Class Lever"

"First-Class Lever (Force and Distance Measurement)"

"Second-Class Lever"

"Third-Class Lever"

Wheel and Axle
Page 57

The wheel and axle is really a form of the lever. When the wheel or the axle turns, the other turns also. The wheel and axle help us turn something more easily or move something across a surface more easily.

Gears are wheels that turn together. If the wheels are different sizes, one turns faster. The number of times the driving wheel turns for each turn of the wheel being driven is called the gear ratio. A ratio of 3:1 means the driven wheel turns once for every three turns of the driving wheel, but with more force.

On the "Wheel and Axle" activity sheet, students are given directions for making a simple wheel and axle. By doing a simple activity it becomes obvious that using a wheel and axle makes forward movement easier and faster.

The Wedge
Pages 58 and 59

A wedge is like two inclined planes back to back. While something moves **over** an inclined plane, a wedge moves **through** something. It is used to split, cut, or go through something. A wedge moves into or under an object.

The two simple activities on "The Wedge" and "Using a Wedge" demonstrate how a wedge can be used to go through wood and how it can be used to separate two objects that are bound together tightly.

The Screw
Pages 60 and 61

The screw is a form of inclined plane that winds around in a spiral. The line that goes around is called the *thread*. A screw is used to hold things together.

The "What Is a Screw" activity sheet examines how a screw is made and how it resembles an inclined plane.

"As the Thread Turns" compares screws with different amounts of thread to see which is easier to screw into something. It also provides an activity to test the effect of using a lubricant on the screw.

Pulleys
Pages 62–64

When you use a fixed pulley, the load moves up and down but the pulley is fixed to something. A fixed pulley reverses the direction of the force. As force is applied downward, the load moves upward. A movable pulley is attached to the load and moves up and down with the load. It reduces the amount of force required to lift a load. By combining a fixed pulley and a movable pulley, a load can be lifted with half as much force (because of the movable pulley) and the load can be moved upward by pulling downward. Multi-pulley systems have the benefit of reducing the strain on the cable, but adding more pulleys to the hook end will add to the load being lifted.

PULLEYS

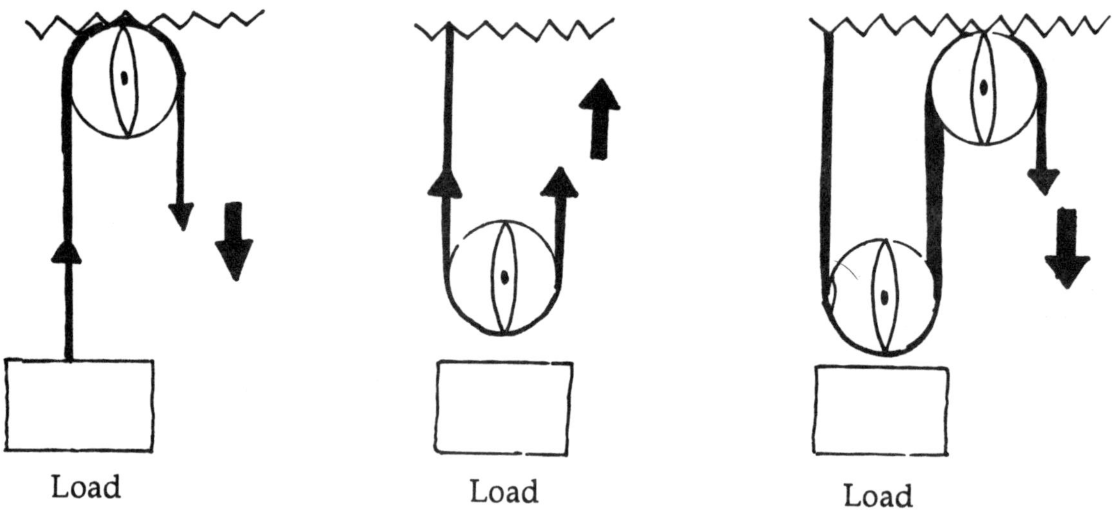

Fixed Single Pulley

Movable Single Pulley

2:1 Advantage

Movable and Fixed Pulley

2:1 Advantage

Three activity sheets: "Fixed Pulley," "Movable Pulley," and "Combining Fixed and Movable Pulleys" are provided to give students activities in which they can work with different types of pulley systems.

Name ______________________________

BOOK LIFT

MATERIALS: 3 books tied together
Smooth, flat board about ½" x 10" x 6"
String
Spring scale
Stack of seven books or something of equal height on which to prop board

DIRECTIONS:

1. Use the board and stack of seven books to make an inclined plane.
2. Pull the three tied books up the board with a spring scale. Record the weight on the spring scale.
3. Lift the three tied books with a spring scale to the same height as the stack of books without using the board. Record the weight on the spring scale.

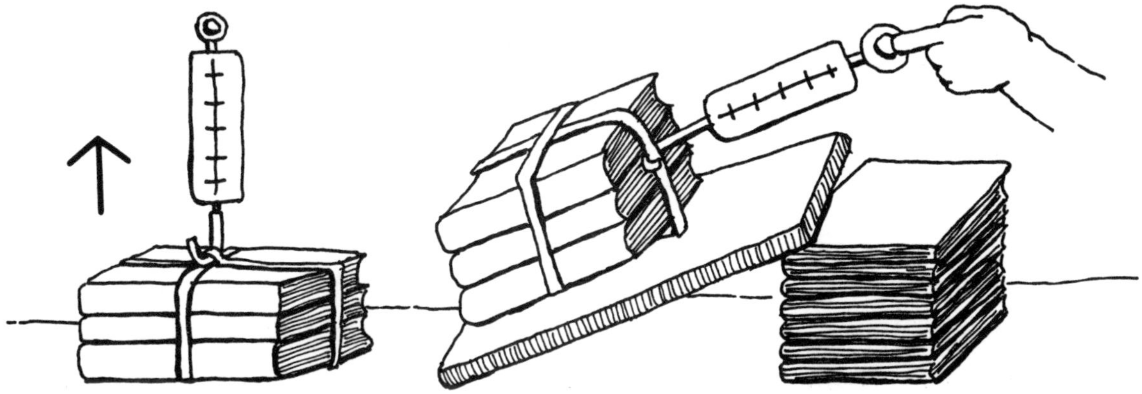

QUESTIONS:

1. Which way was it easier to lift the books?

____

2. What would happen if you used a higher stack of books? Try it and find out. Tell what happened.

____

3. Try the following variations. Record your results on the activity sheet, "Friction Fighters."
   - Use rounded pencils under the books as rollers.
   - Cover the board with wax paper, sandpaper, and a heavy textured fabric.
   - Place books on a roller skate.
   - Place large rubber bands around the board about 2" apart.
   - What other materials can you think of to use to reduce friction? Write them down in spaces 8, 9, and 10 on the chart and test them.

Name ____________________

FRICTION FIGHTERS

	Variations	Spring Scale Reading
	1. Pulling books up board.	
	2. Using pencils as rollers.	
	3. Covering board with wax paper.	
	4. Covering board with sandpaper.	
	5. Covering board with heavy textured fabric.	
	6. Placing books on roller skates.	
	7. Covering board with rubber bands.	
YOUR IDEAS	8.	
	9.	
	10.	

QUESTION:

What conclusions can you draw about the materials which best reduce friction?

____

____

____

Name ______________________________

LIFTING LOADS WITH A FIRST-CLASS LEVER

MATERIALS: Yardstick
String
Fulcrum (1-4" high)
Spring balance
Several books

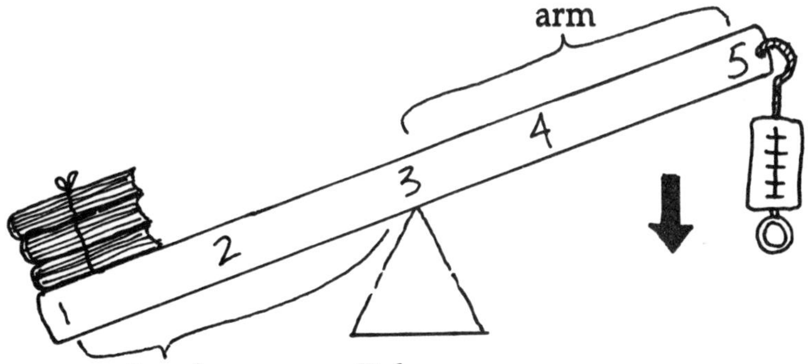

DIRECTIONS:

1. Record all measurements for this activity on the activity sheet, "First-Class Lever (Force Measurements)."
2. Place the fulcrum at position 3.
3. Place three books at position 1.
4. Attach a spring balance to the other end of the lever at position 5 and pull down. Record the force that is required to lift the books to a level position.
5. Repeat with the fulcrum at position 4, 2, between 2 and 3, and between 3 and 4. Record the force measurements on the activity sheet.
6. Measure the length of the effort arm at each fulcrum position. Record measurements.

QUESTIONS:

1. Was the load easier to lift when it was closer to or farther from the fulcrum?

   ____

2. When the fulcrum is closer to the load, is the effort arm longer or shorter?

   ____

3. After comparing the force measurements listed on the activity sheet, what conclusion can you draw about length of the effort arm and the force required to do work?

   ____

   ____

   ____

Name ______________________________

FIRST-CLASS LEVER (FORCE MEASUREMENTS)

Fulcrum Position	Length of Effort Arm	Force
3		
4		
2		
2 – 3		
3 – 4		

Make a statement about the relationship of the length of the effort arm and the amount of force required to do work.

____

____

____

____

____

____

____

____

Name ______________________________

RELATIONSHIP OF FORCE AND DISTANCE USING A FIRST-CLASS LEVER

MATERIALS: Yardstick
Spring balance
Fulcrum (4" high)
3 books
String
Ruler

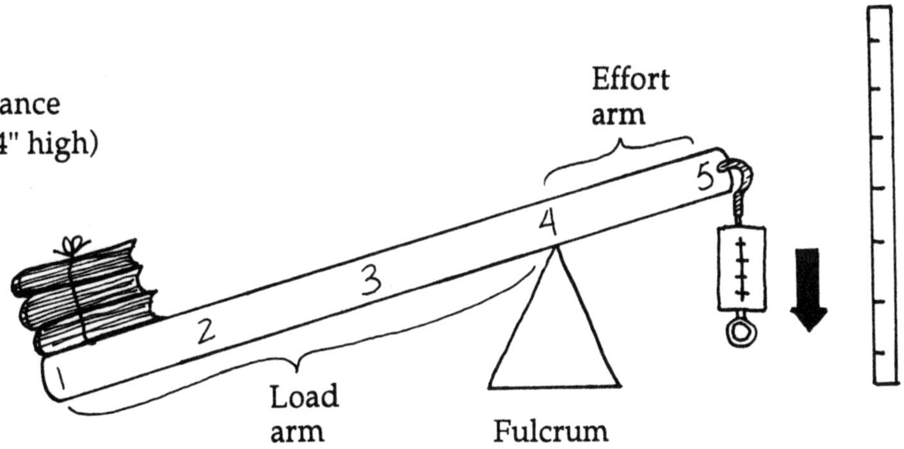

DIRECTIONS:

1. Tie the books together and place at one end of the lever. Attach the spring scale to the other end.
2. Place a measuring stick near the end of the lever as shown.
3. Pull down on the spring balance to lift the books and record the force with the fulcrum at positions 3, 4, 2, between 2 and 3, and between 3 and 4. Record all measurements on the following activity sheet, "First-Class Lever (Force and Distance Measurements)."
4. Record the distance the effort arm moves at each fulcrum position while the load arm is at the distance indicated on the chart.

QUESTIONS:

1. As force decreases, does the distance the effort arm travels increase or decrease?

   ____

2. As the force decreases, does the distance the load arm travels increase or decrease?

   ____

3. Can you make a statement about the relationship of force and the distance the load travels, and the distance the effort arm travels as the fulcrum gets closer to the load?

   ____

   ____

   ____

Name ______________________________

FIRST-CLASS LEVER (FORCE AND DISTANCE MEASUREMENTS)

Fulcrum Position	Force	Distance Load Arm Traveled	Distance Effort Arm Traveled
3		10 cm	
4		5 cm	
2		5 cm	
2 – 3		5 cm	
3 – 4		10 cm	

Make a statement about the relationship between force and the distance the load travels and the distance the effort arm travels as the fulcrum gets closer to the load.

____

____

____

____

____

Name ____________________

SECOND-CLASS LEVER

(In a second-class lever both the load and the effort travel in the same direction.)

MATERIALS: Yardstick
Fulcrum
3 books
String

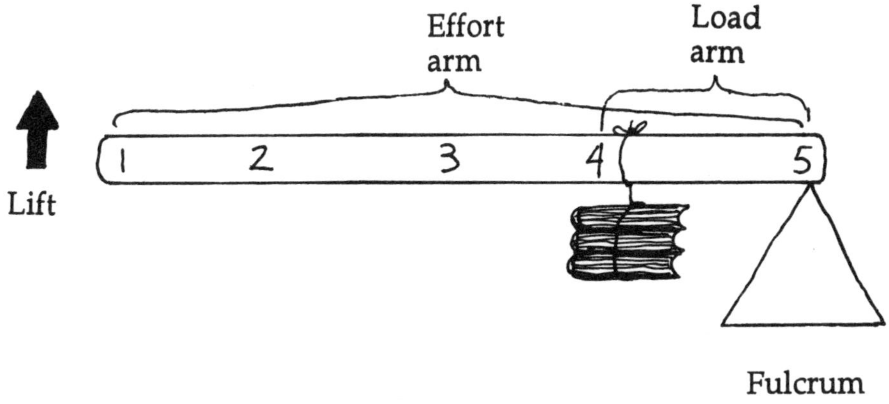

DIRECTIONS:

1. Place books at position 4 and lift the lever.
2. Try lifting with the books at positions 3, 2, and 1.
3. With the books at position 4, measure the distance of the effort arm and the load arm.
   Effort arm ____________________
   Load arm ____________________
4. With the books at position 2, measure the distance of the effort arm and the load arm.
   Effort arm ____________________
   Load arm ____________________

QUESTIONS:

1. Is it easier to lift books when they are closer to or farther from the fulcrum?
   ____
2. Is it easier to lift books when the effort arm is longer or shorter?
   ____
3. Can you make a statement about the relationship of force, the length of effort arm, and the length of the load arm?
   ____
   ____
   ____
   ____

Name ____________________

THIRD-CLASS LEVER

MATERIALS: Yardstick
3 books
String
Table

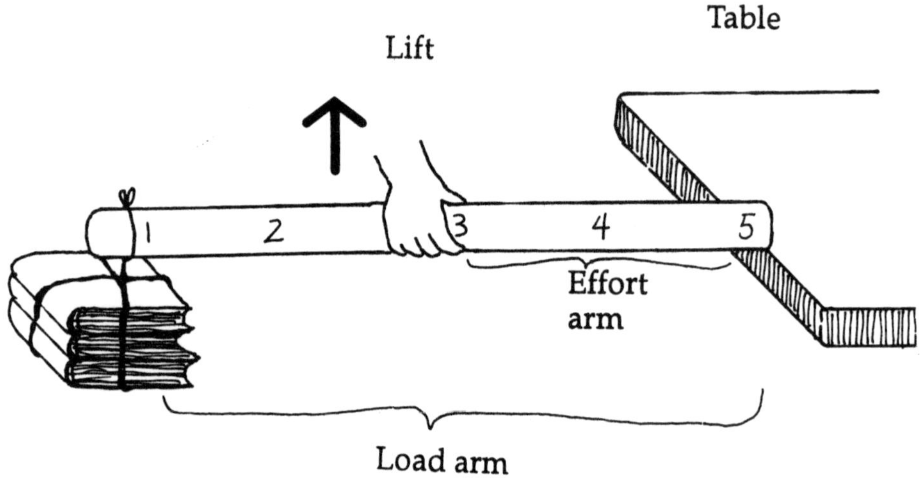

DIRECTIONS:

1. Use the edge of the table as a fulcrum.
2. Suspend the books at position 1.
3. Holding the lever at position 3, lift the books.
4. Next lift the load at position 2 and position 4.

QUESTIONS:

1. Is it easier to lift the books with your hand closer to or farther from the fulcrum?

    ____

    ____

2. How is a third-class lever different from a first and second-class lever?

    ____

    ____

    ____

    ____

    ____

    ____

Name ______________________________

WHEEL AND AXLE

MATERIALS: Cardboard circle (10 cm in diameter)
Pencil
Tape measure

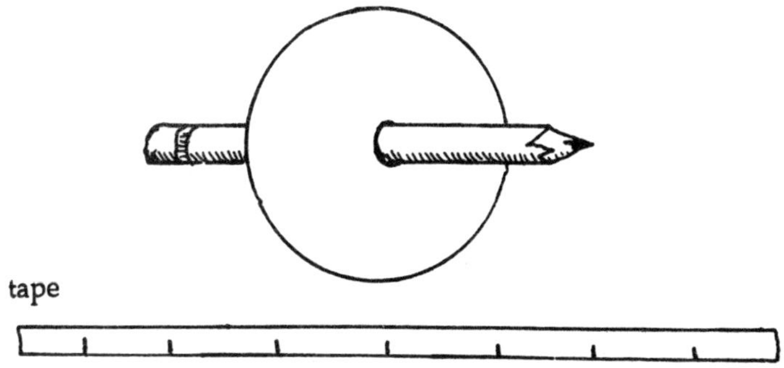

DIRECTIONS:

1. Insert the point of the pencil through the center of the circle. This is a wheel and axle.
2. Measure the distance the wheel travels in one rotation.

   ____

3. Remove the pencil from the cardboard circle and measure how far it moves in one rotation.

   ____

4. Measure how far the pencil would travel in 10 rotations.

   ____

5. Now put the pencil (axle) back into the cardboard circle (wheel) and see how far it travels in 10 rotations.

   ____

QUESTIONS:

1. Can you name an advantage of the wheel and axle?

   ____

   ____

2. Give some examples of the use of the wheel and axle.

   ____

   ____

   ____

Name ____________________

THE WEDGE

MATERIALS: Large nails with and without points
Hammers
Thick board

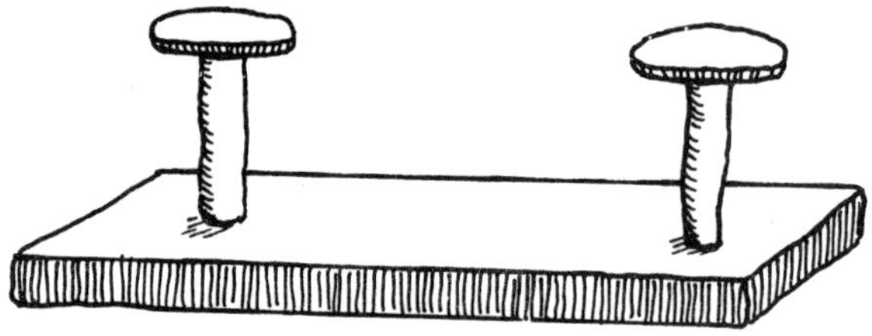

DIRECTIONS:

1. Measure the nails. Record the results.
2. Hammer one of the large nails with no point into the board 20 times. Measure how much of the nail is still out of the board.
3. Hammer one of the nails with a point into the board 20 times. Measure how much of the nail is still out of the board.

	Nail without point (wedge)	Nail with point (wedge)
size before hammering	____________	____________
size after hammering	____________	____________

QUESTION:

Which nail was easier to pound? Explain why.

____
____
____
____
____
____
____
____

Name ______________________________

USING A WEDGE

MATERIALS: 2 wood blocks
2 large, thick rubber bands
1 wedge

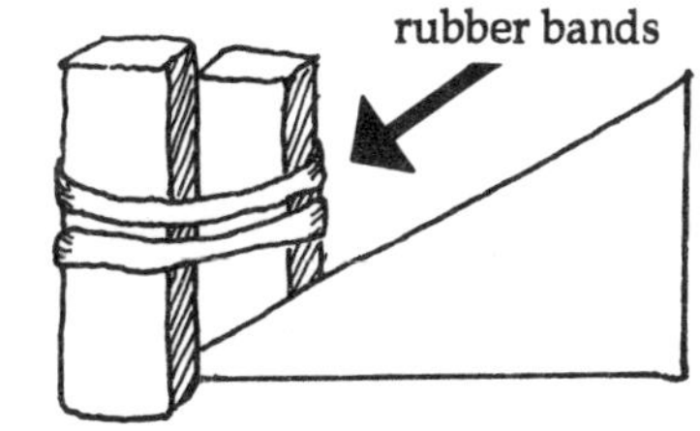

DIRECTIONS:

1. Wrap each rubber band twice around the wood blocks to hold them together tightly.
2. Try to pull the blocks apart with your fingers.
3. Put the sharp edge of the wedge where the surfaces of the blocks come together and press the wedge between the blocks, trying to separate them.

QUESTIONS:

1. How did the wedge help in separating the blocks?

____

____

2. What are some other wedges that are used every day? Explain their uses.

____

____

____

____

____

____

____

____

Name ____________________

WHAT IS A SCREW?

MATERIALS: Large pencil
Sheet of white paper
Scissors
Black marker
Large screw

DIRECTIONS:

1. Cut the piece of paper diagonally from corner to corner making a triangle. Make a heavy black line with the marker along the longest edge of the triangle.

Cut

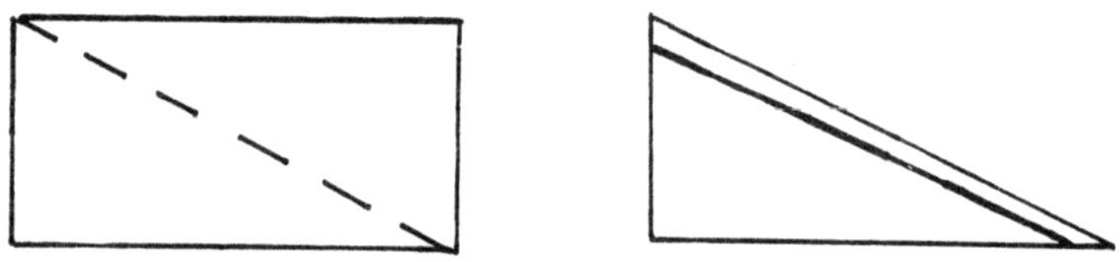

2. Hold the triangle upright on your desk and, beginning with the shortest side, wrap the triangle around your pencil. Use tape to hold the paper if needed.
3. Hold the wrapped pencil next to the large screw. Compare the heavy line on the triangle with the threads on the screw.

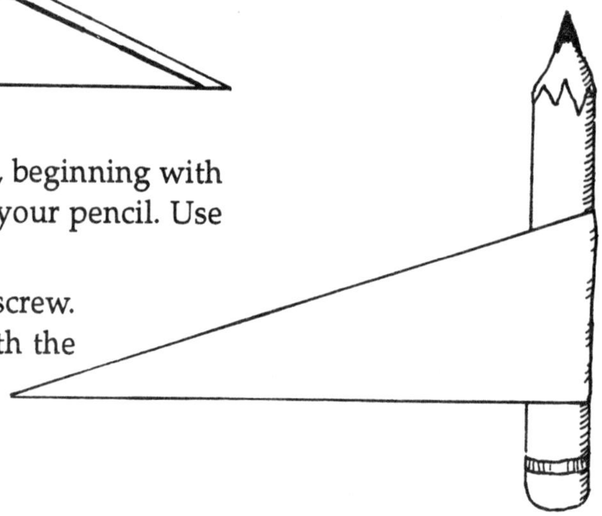

QUESTIONS:

1. Do the heavy lines on the triangle resemble the threads on the screw? Explain.

____

____

2. Think about how the black edge of the triangle looked before it was wrapped. How do the threads of the screw compare to an inclined plane?

____

____

3. Which distance is greater, the path straight up the side of the pencil or the path spiraling around it? Measure both and record your answers below.

Path straight up the side of the pencil ____________________ inches

Path spiraling around the pencil ____________________ inches

Name ______________________________

AS THE THREAD TURNS

MATERIALS: Wood block
Two screws the same size but different thread
Screwdriver
Masking tape
Lubricant (oil or petroleum jelly)
Nail the same size as screw

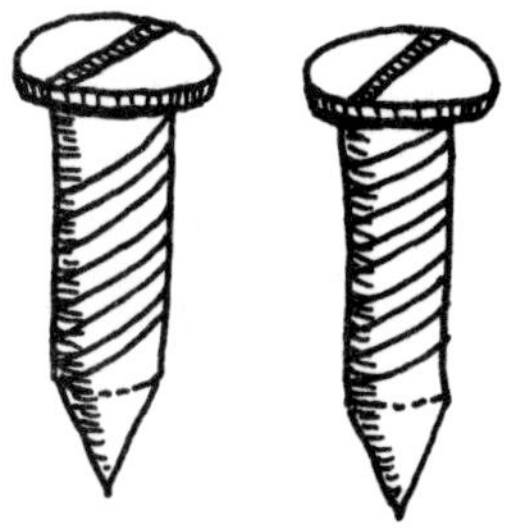

DIRECTIONS:

1. Place a piece of masking tape around the handle of the screwdriver. Make a dark mark on the tape. Each time the mark comes back to the place it started, you will count one turn.
2. Count how many turns it takes to screw the screw with the least thread all the way into the wood. Record the results.
3. Repeat with the other screw.
   Screw with the least thread____________ turns.
   Screw with the most thread ____________ turns.
4. Put some type of lubricant (oil or petroleum jelly) on the screw. Does this affect the number of turns? Does this have any effect on effort? Record your results.
   Screw with least thread and oil ___________ turns.
   Screw with most thread and oil ___________ turns.
   Was it easier or harder using oil? ______________________________
5. Hammer a nail the same size as your screw into the block of wood until it is all the way into the wood. Was it easier to hammer the nail or screw into the wood?

QUESTIONS:

1. Why did it take more turns for the screw with the most thread to go in? (Hint: Compare the depth of the holes that were drilled by each screw.)

   ____

   ____

2. Which requires more effort, the screw with more or less threads? Explain your answer.

   ____

   ____

Name ______________________________

FIXED PULLEY

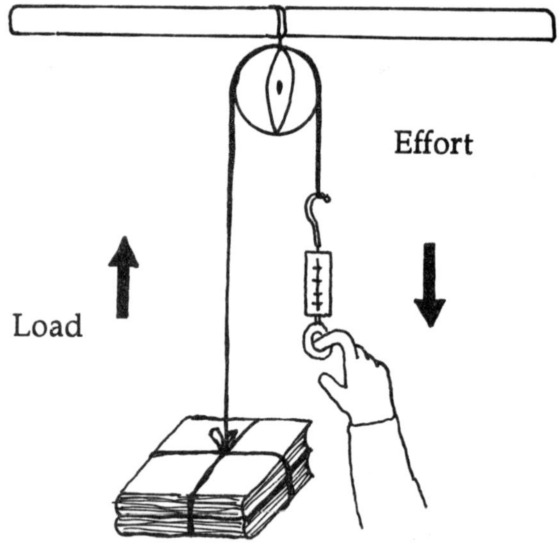

MATERIALS: Single-wheel pulley
Crossbar
Spring balance
Heavy string
Pencil
2 books

DIRECTIONS:

1. Arrange materials as shown in the illustration.
2. Pull down on spring balance to lift the books 20 cm. How much force was used to lift the books?

   ____
3. Now lift the books using only the spring balance. How much force was used to lift the books this time?

   ____
4. As you pull down (force), in what direction do the books travel?

   ____

QUESTIONS:

1. Did using the pulley decrease the amount of force needed to lift the books?

   ____

   ____
2. Did using the pulley make the books move in the opposite direction from the force or effort?

   ____

   ____
3. What is the advantage of using a fixed pulley?

   ____

   ____

Name ____________________

MOVABLE PULLEY

MATERIALS: Pulley
Crossbar
Spring balance
Heavy string
Pencil
2 books

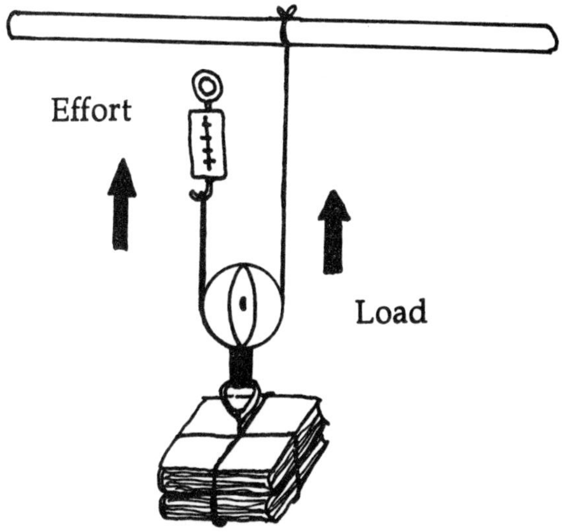

DIRECTIONS:

1. Weigh the books with the spring balance.

   ____

2. Arrange materials as shown in the illustration above.
3. Lift the books by pulling up on the spring balance. Record the amount of force used to raise the books steadily.

   ____

4. Compare the weight of the books in #1 with the force required to lift the books in #3. Which is greater?

   ____

QUESTIONS:

1. Did the pulley decrease the amount of force needed to lift the books?

   ____

2. Did the pulley cause the load to move in the opposite direction of the force or did the load move in the same direction?

   ____

3. Make a statement telling what effect a movable pulley has on the force required to lift a load and what effect it has on the direction the load travels compared to the direction of the effort or force.

   ____

   ____

   ____

Name ____________________

COMBINING FIXED AND MOVABLE PULLEYS

MATERIALS: 2 pulleys
Crossbar
Spring balance
Heavy string
Pencil
2 books

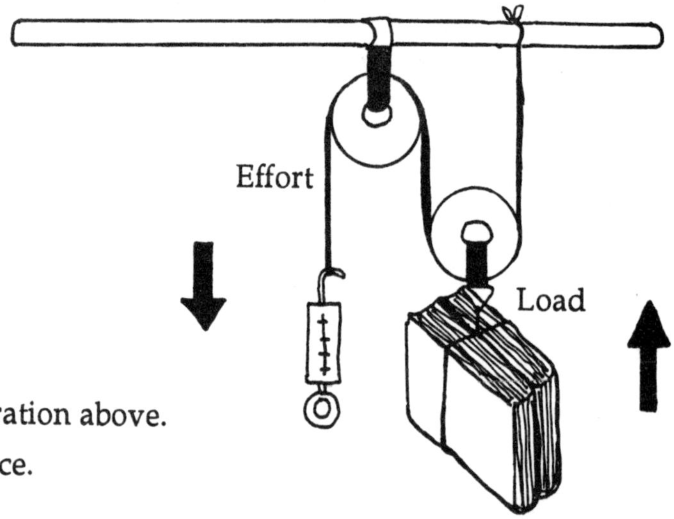

DIRECTIONS:

1. Set up materials as shown in the illustration above.
2. Weigh the books with the spring balance.

   ____
3. Pull down in the effort position with the spring balance. Record the force required to lift the books steadily.

   ____
4. What direction do the books move as you pull down at the effort position?

   ____

QUESTIONS:

1. Did the pulley system used decrease the amount of force needed to lift the load (books)?

   ____
2. What direction did the load move compared to the direction of the force or effort?

   ____
3. What are the advantages of a combined pulley system?

   ____

   ____

Try This!

Experiment with different combinations of pulleys and test the mechanical advantage of your system. Illustrate your system and explain its advantages.

CHAPTER 5 Matter

TEACHER NOTES

Matter Changing State, Liquid to Solid Pages 69 and 70

Matter is anything that takes up space. Matter can exist in three different states—solid, liquid, or gas. States of matter depend on how the particles inside the matter are linked. In a solid, the particles are linked tightly together therefore it holds its shape. In a liquid the particles are held together less tightly so they can slide past one another, enabling a liquid to flow. Liquids have volume and they take the shape of the containers in which they are placed. Particles in a gas move about freely in all directions and are not held together tightly. The container holding the gas determines its volume and shape. Gas does not have a volume nor shape of its own. The higher the temperature, the more quickly and freely the particles move. Therefore, temperature and pressure can change the state of matter.

"Matter Changing State" is a simple activity dealing with the evaporation of water—liquid (water) changing to a gas (water vapor).

The activity in "Liquid to Solid" asks students to try converting melted butter and water into solids by cooling them. Students are asked to compare what happens. Water, an exception to the rule, expands when it becomes solid! The butter will shrink or contract as it solidifies.

Making Carbon Dioxide Page 71

Gas takes up more space than solids and liquids because the molecules are spread further apart. In this activity, vinegar and bicarbonate of soda combine and produce carbon dioxide. The gas given off will inflate the balloon.

Viscosity of Liquids Page 72

Viscosity refers to a liquid's thickness or its resistance to flow. In this activity, students drop a marble into various liquids and calculate the time it takes the marble to reach the bottom. The viscosity is compared in various liquids. Encourage students to do some experiments of their own, using liquids not used in this activity.

Measuring Volume (Solids and Liquids) Pages 73 and 74

Fluid volume is measured in milliliters and liters. The cubic centimeter or meter is used to measure the volume of solids. (volume = length x width x height)

Using liter containers, students will experiment measuring liquid volume. In this activity, they will also have an opportunity to calculate the volume of various solids by using the formula V = L x W x H. In the last activity, they are asked to figure out how to calculate the volume of a solid using water displacement.

Molecules Meshing Together Page 75

Scientists believe the atom is the smallest particle of an element that still retains the properties of the element. The atom can be thought of as electrons orbiting around a nucleus of protons and neutrons, even though the atom has proved to be much more complicated. Atoms have been smashed to show the existence of scores of different particles and energy packages called *quanta*.

Molecules are pairs of atoms that have joined together. The molecular structure of some substances like iron, steel, and cobalt make them magnetic.

Elements, or basic, pure chemicals can be combined to make *compounds*. Each element has its own characteristics and can exist as a solid, liquid, or gas. All atoms of an element are the same but different from the atoms of any other element. Compounds are formed when the atom or atoms of one element join with those of another element. The properties of a compound are usually different from those of the elements that make up the compound. Most matter on earth are compounds. One hundred nine different elements have been discovered so far. Their characteristics depend upon the makeup of their electron shells. Elements have been grouped by being written down in order of their atomic numbers (the number of protons in the nucleus). By looking at an element's position in the periodic table, scientists can tell a lot about how the element will behave.

In "Molecules Meshing Together" students try to determine why one cup of water and one cup of alcohol do not equal two cups of liquid!

Goop Page 76

"Goop" is an example of chemical change and interaction of substances to form a new compound. In this activity, students experiment with the new compound, Goop, and study its characteristics.

Rust
Page 77

When there is a physical change, only the state (solid, liquid, or gas) is changed and the change is almost always reversible. When a chemical change occurs, at least one of the elements or compounds is changed and new compounds are formed. There is a change in molecular structure and the change is not reversible. Ice melting is an example of a physical change. Burning or something rusting are examples of a chemical change. When something burns, it joins with oxygen in the air and produces heat, water, and carbon or carbon dioxide.

Experimenting with rust, students find out which liquids enhance or inhibit the formation of rust when iron and oxygen combine.

Testing for Acids and Bases
Page 78

All substances that are *acid* contain hydrogen. Mild acids like vinegar and lemon juice taste sour. A strong acid, like sulfuric acid, can burn and even dissolve metals.

A *base* is the opposite of an acid. A weak base like baking powder tastes bitter and feels soapy. Strong bases, such as caustic soda, can be as dangerous as strong acids. A base that dissolves in water is called an *alkali*. When an alkali is added to an acid, it cancels out the acidity, neutralizing it.

In this activity, various substances are tested using an indicator solution made by simmering cabbage leaves or by using a homemade litmus paper. Substances tested are then categorized as acidic or basic.

Mixtures Versus Solutions
Page 79

A *mixture* is when two or more substances are mixed together but retain their separate identities. Elements and compounds in a mixture can often be separated. In a solution, the substance placed in the liquid becomes part of the liquid. In other words, the molecules of the solid (the solute) are completely mixed with the molecules of the liquid (the solvent).

Experimenting With Dissolving Pages 80–84

In a solution, the substances that mix together are said to be dissolved. In the "Experimenting with Dissolving" activity sheet, students experiment with different solvents as well as using different temperatures, stirring, and changing the surface area of the solute to see what effects these changes have on the rate of dissolving. Charts to record findings for each of these activities are included.

Miscible and Immiscible Liquids Page 85

Liquids that dissolve in each other are *miscible*. Liquids that do not dissolve in each other are *immiscible*. In this activity, students experiment with water, oil, and food coloring to see how each reacts with the other when mixed.

Growing Crystals Charcoal Gardens Pages 86–89

If a solution becomes saturated, no more solutes will dissolve in the liquid. As the solution cools down or evaporates, it becomes more than saturated and the solute molecules begin to link up with one another and grow into solid crystals.

No two crystals are alike in appearance. The rate of cooling or evaporation affects the size of the crystal. Slow cooling forms larger crystals.

Almost every pure, solid substance is crystalline. All rocks and metals are crystals, although the crystals are too small to be visible.

Both activity sheets "Charcoal Gardens" and "Growing Crystals" and their accompanying charts for recording data and observations deal with various methods of growing crystals.

The "Growing Crystals" activity sheet explains the effects of rate of cooling, space, and concentration of the solution on the size and shape of crystals.

In "Charcoal Gardens" students try changing the container, changing the charcoal to another material, and changing the temperature to see if these things affect the growth of crystals.

Name ______________________________

MATTER CHANGING STATE

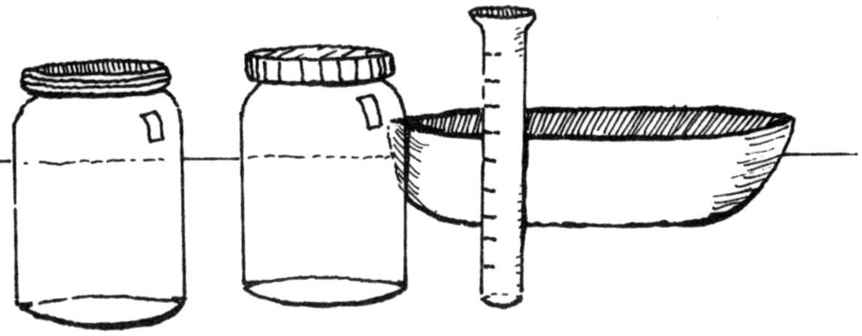

MATERIALS: 2 jars, one with a lid
Large shallow bowl
Graduated cylinder
Water

DIRECTIONS:

1. Add 200 ML water to each jar and the shallow bowl.
2. Place a lid on one of the jars.
3. After three or more days, measure the amount of water left in each jar and in the bowl. Record your results on the chart.

Container	Water after 3 days
Jar	
Jar with Lid	
Shallow Bowl	

QUESTIONS:

1. Under what conditions does liquid water change to water vapor the fastest?

____

____

2. Would water in a large shallow puddle or a small deep puddle dry up the fastest? Explain your answer.

____

____

____

Name ______________________

LIQUID TO SOLID

When liquids cool they become solid. Their atoms and molecules are closer together and take up less space than in a liquid. Most substances shrink when they solidify.

DIRECTIONS:

Pour some melted butter into a small glass jar. Fill the jar to the brim and refrigerate until solid. Look at the jar. What do you see? Can you explain what happened?

____

____

____

____

____

____

Try This!

(Unlike other substances, water expands when it freezes and becomes solid.)

Fill a plastic soda bottle with water and place it in the freezer. Describe what happens when the water freezes.

____

____

____

____

____

____

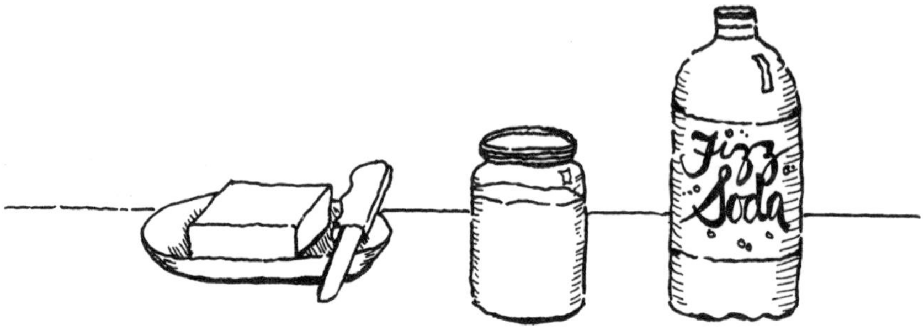

Name ______________________________

MAKING CARBON DIOXIDE

(Gas takes up more space than solids or liquids)

MATERIALS: Vinegar
Bicarbonate of soda
Balloon
Narrow-necked bottle

DIRECTIONS:

1. Pour vinegar into the bottle until it is about one-fourth full. If the vinegar is warm, the reaction will be faster.
2. Pour bicarbonate of soda into the neck of the balloon until the ball of the balloon is filled with soda.
3. Keeping the ball of the balloon hanging down, stretch the neck of the balloon over the neck of the bottle.
4. Lift the balloon quickly so the soda falls into the bottle. Shake the bottle to be certain the vinegar and soda are well mixed.
5. Observe what happens to the balloon. What gas is given off when the vinegar and soda are mixed? Explain why the balloon inflates.

____

____

____

____

1 2, 3 4, 5

Name ______________________

VISCOSITY OF LIQUIDS

(Viscosity is the thickness or resistance to flow.)

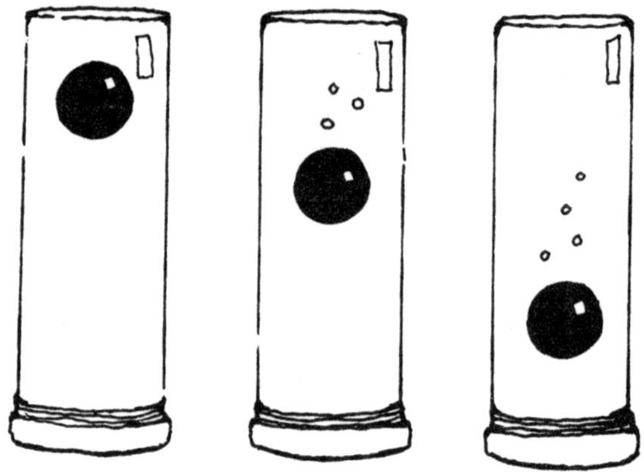

MATERIALS: 6 tall jars (olive jars)
6 marbles
Corn syrup
Water
Milk
Cooking oil
Glycerol
Wine

DIRECTIONS:

1. Be sure all jars are the same size. Place a marble in each jar.
2. Fill each jar with one of the liquids above. Put the lids on tightly.
3. Get someone to help and turn all of the jars over at the same time. Observe the marbles.
4. Record which marble sinks to the bottom first, second, third, etc. List the substances in order of their viscosity.

   1.______________ 4.______________
   2.______________ 5.______________
   3.______________ 6.______________

5. Experiment with some other liquids. Try timing how long it takes for the marble to sink to the bottom of each jar. Record your results on a chart.

Liquid	Time for Marble to Fall

Name ____________________________

MEASURING VOLUME (SOLIDS AND LIQUIDS)

Volume = length x width x height (V = L x W x H)
The cubic centimeter or meter is the unit used for measuring the volume of a solid.
Liters and milliliters are used to measure liquid volume.

MATERIALS: 10 jars (assorted sizes)
Cans (assorted sizes)
Graduated cylinder
Different size blocks of wood
Water
10 bottles (assorted sizes)
Cartons (assorted sizes)
Square and rectangular-shaped boxes
Small rock

DIRECTIONS:

1. Use a liter-size plastic drink bottle to guess the volume of each container (jars, bottles, cans, and cartons).
2. List the containers according to their volume.

Less than 1 Liter	1 Liter	More than 1 Liter

3. Measure each of the boxes and blocks of wood. Calculate their volume by using the formula given above. Your answer should be given in cubic centimeters.

Boxes and Wood Blocks	Measurement length, width, height	Volume
a.		
b.		
c.		
d.		
e.		

Name ____________________

MEASURING VOLUME (SOLIDS AND LIQUIDS)

4. Fill a graduated cylinder with some water. Record the volume of the water.
   Drop a small rock into the water and measure the volume again.
   Can you figure out the volume of the rock?
   Volume of the water ____________________
   Volume of the water and rock ____________________
   Volume of the rock ____________________

QUESTIONS:

1. What is the formula for finding the volume of a solid?

   ____

2. Explain how you would use water in a graduated cylinder to find the volume of a solid object.

   ____

3. By using the method you explained in question 2, find the volume of two marbles. Show all of your work.

   ____

Name ________________________________

MOLECULES MESHING TOGETHER

MATERIALS: 1 c. water
1 c. rubbing alcohol
1 two-cup measuring cup

DIRECTIONS:

1. Predict how much liquid you will have if you pour 1 c. water and 1 c. alcohol together.

   My guess: 1 c. water + 1 c. alcohol = ____________ c. liquid.

2. Pour 1 c. water and 1 c. alcohol into the two-cup measuring cup. Measure how much liquid you have. Write your answer here:

   1 c. water + 1 c. alcohol = ____________ c. liquid.

QUESTION:

Was the amount of liquid more or less than two cups when poured together? Can you explain why? (The title may give you a clue!)

____

____

____

____

____

____

? ? MYSTERIOUS MIXTURE ? ?

Mix iron filings and sugar together. Can this mixture be separated? Use magnets, water, spoons, screens or a strainer, and jars to see if you can separate the things in this mixture.

Explain below what you did to separate the iron filings and the sugar.

____

____

____

Name ______________________________

"GOOP"

MATERIALS:

Borax	White glue	Water
Paper cups	30 ML cups	Lysol
Food coloring	Zip-loc bags	Paper towels
Stirring sticks		

DIRECTIONS:

1. Ahead of time mix ¼ c. of Borax with 1 qt. water. Let the mixture settle and use only the clear portion.
2. Mix 15 ML (1T.) of white glue with 15 ML of water. Add a few drops of food coloring if desired. Pour the mixture into a paper cup and stir. Add a few drops of Lysol. The Lysol inhibits spoiling.
3. Pour 10 ML (2t) of the Borax mixture (#1) into the thinned glue and stir quickly. Knead the mixture with your hands.

Experiment and answer these QUESTIONS:

1. What is the consistency of "goop"?

   ____

2. What happens when "goop" is agitated?

   ____

3. Will "goop" stretch? Explain.

   ____

   ____

4. If you make a ball with "goop" will it bounce?

   ____

5. If you put "goop" on the end of a straw, can you blow it up?

   ____

Think up an experiment you can try with "goop."

Name ______________________________

RUST

(Iron + Oxygen = Rust)

MATERIALS: Steel wool
Vinegar
Water
3 pie pans

DIRECTIONS:

1. Place a steel wool pad in each pan.
2. Pour 25 ML water over the steel wool pad in pan 1.
   Pour 25 ML vinegar over the steel wool pad in pan 2.
   Leave the steel wool pad in pan 3 dry.
3. Leave the steel wool pads in the pans for several days. Check the steel wool for rust each day and record your observations.

	Day 1	**Day 2**	**Day 3**	**Day 4**	**Day 5**
Water					
Vinegar					
Air					

QUESTIONS:

1. In which pan did the steel wool rust the fastest?
   ____
2. What conclusions can you draw about things rusting in water vs. vinegar?
   ____
3. What conclusions can you draw about rusting in water vs. air? In vinegar vs. air?
   ____

Things to Try!

Experiment by putting steel wool in some other liquids—laundry detergent, dish washing liquid, tea, oil, or apple juice. Make a chart to record your results.

Name ______________________________

TESTING FOR ACIDS AND BASES

MATERIALS:

1 head of raw, red cabbage	Small pan	Large jar
Strainer	Small jars	Baking soda
Vinegar	Liquid soap	Salt
Lemon juice	Orange juice	Carrot
Soda pop	Tomato	Milk
Assorted vegetables		

Substance	Acid	Base
Liquid Soap		
Salt		
Lemon Juice		
Orange Juice		
Carrot		
Soda Pop		
Tomato		
Milk		

DIRECTIONS:

1. First make indicator solution. Let cabbage leaves simmer in a saucepan for 15 minutes, and strain the water into a large jar. To test, put 1 tsp. baking soda into a small jar. Put 1 tsp. vinegar into another small jar. Add some cabbage water to each jar. Baking soda is a base and should turn the water green. Vinegar, an acid, should turn the water red.

   Make litmus paper by soaking newspaper strips (taken from the borders where there is no print) in cabbage water and then letting them dry.

2. Use the cabbage water or litmus paper to test the materials on the list above to determine whether they are acids or bases. Can you find some other things to test?

Acid ➔ RED
Base ➔ GREEN

Name ______________________________

MIXTURES VERSUS SOLUTIONS

MATERIALS: 2 glass jars
Spoons for stirring
Water
Sugar
Gravel or small stones
Marbles
Paper clips
Toothpicks

DIRECTIONS:

1. Fill each glass half full of water.
2. Put gravel, marbles, toothpicks, and paper clips in one jar.
3. Put a few tablespoons of sugar in the second jar.
4. Stir each jar. Compare the results.

QUESTION:

Which of the above is a mixture and which is a solution? Explain your answer.

____

____

Try This!

Mix each of the following in some water to see if they produce a mixture or a solution. Check the answer.

Substance	Mixture	Solution
Sand		
Powdered Milk		
Powdered Chocolate		
Cornstarch		
Instant Coffee		
Other		

Name ______________________________

EXPERIMENTING WITH DISSOLVING

MATERIALS: Hard candy
Aspirin
Water
Soap powder
Soda
Salt
Milk
Sugar
Oil
Cornstarch
Liquid detergent
5 small jars

DIRECTIONS:

Using Different Liquids (Solvents)

1. Place a piece of hard candy in your mouth. Without chewing it, see how long the candy takes to dissolve. Record your results on the "Using Different Liquids" chart.
2. Place a piece of candy in each of the liquids above. Does the candy dissolve? How long did it take? Record your results on the "Using Different Liquids" chart.

Different Temperatures

3. Next place a piece of candy in a jar of cold water. Place another piece of candy in a jar of warm water and a third piece of candy in a jar of hot water. Use the "Different Temperatures and Stirring" chart to record how long it takes the candy in each jar to dissolve.

Stirring

4. Place a piece of candy in each of two jars half filled with warm water. Stir the candy in one jar. Do not stir the other jar. Does stirring help the candy to dissolve faster? Record your results on the "Different Temperatures and Stirring" chart.

Surface Area

5. Fill five jars with a little warm water. In one jar, put a whole piece of hard candy. In the next jar, break the candy in half. In the third jar, break the candy into four pieces. In the fourth jar, break the candy into eight pieces, and finally in the last jar, crush the candy so that it is pulverized. How long does it take for the candy in each jar to dissolve? Record your results on the "Surface Area and Dissolving" chart.

Name ______________________________

Using Different Solutes

6. Try dissolving some common household items in water. Try dissolving aspirin, soap powder, salt, sugar, cornstarch, and any other items you can name. Record the results on the chart below.

Household Items	Dissolved	Not Dissolved
Aspirin		
Soap Powder		
Salt		
Sugar		
Cornstarch		

List all the household items you tested that will dissolve in water.

Name ____________________

USING DIFFERENT LIQUIDS (EXPERIMENTING WITH DISSOLVING)

Liquid	Not Dissolved	Dissolved	Time
Saliva			
Soda			
Milk			
Oil			
Liquid Detergent			

QUESTION:

Does the type of liquid used make any difference in whether something will be dissolved? Does the type of liquid affect the time it takes the candy to dissolve? Explain your answer.

Name ______________________________

DIFFERENT TEMPERATURES AND STIRRING (EXPERIMENTING WITH DISSOLVING)

	Stirred—Time to Dissolve	Not Stirred — Time to Dissolve
Cold Water		
Warm Water		
Hot Water		

QUESTIONS:

1. Does the temperature of the water affect the rate at which the candy is dissolved? Explain your answer.

____

____

____

____

2. Does stirring affect the rate at which the candy dissolves? Explain.

____

____

____

Name ______________________________

SURFACE AREA AND DISSOLVING (EXPERIMENTING WITH DISSOLVING)

Surface Area	Time to Dissolve
Whole Piece of Candy	
Half Piece of Candy	
Quartered Piece of Candy	
Piece of Candy divided into eighths	
Crushed Candy	

QUESTION:

Does the amount of exposed surface area affect the rate at which something dissolves? Explain your answer using the results recorded on this chart.

____

____

____

____

____

____

____

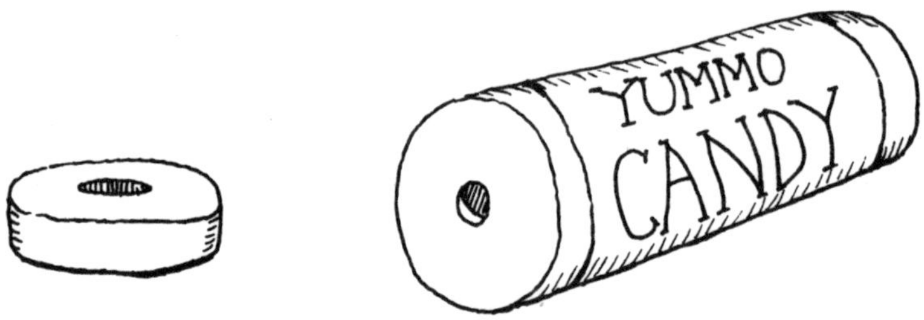

Name ______________________________

MISCIBLE AND IMMISCIBLE LIQUIDS

(Liquids that dissolve in one another are miscible; those that do not dissolve in one another are immiscible.)

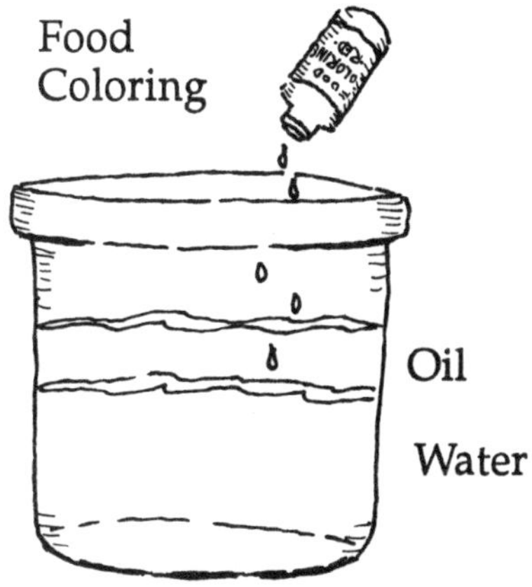

MATERIALS: Cooking oil
Water
Food coloring
Dropper
Glass beaker or large jar

DIRECTIONS:

1. Pour some water into the beaker. Then pour some of the oil on top of the water. Stir the water and oil. Explain what happens. Do you know why?

____

____

____

____

2. Using the dropper, drop several drops of food coloring into the oil. Drop the food coloring in very carefully. What happens to the food coloring in the oil? Can you explain why?

____

____

____

____

3. Using the end of a spoon, push the balls of food color down into the water. What happens when the food color hits the water? Can you explain why?

____

____

____

____

Name ______________________________

GROWING CRYSTALS

MATERIALS: Salt
Sugar
Pencil
String
Small piece of cardboard
Small screw
Shallow plate

1.

DIRECTIONS:

1. Fill a jar with hot water and add salt until it stops dissolving. Prepare a piece of cardboard with a string hanging from it. Thread the string through holes punched in the cardboard. Make the string just long enough to reach the bottom of the jar. Weight the string with a small screw and drop it into the solution. Put the jar into a warm place and let it cool slowly. Observe the salt solution each day and record what happens on the "Growing Crystals" chart.
2. ROCK CANDY — Stir about 2 c. sugar into ¼ c. of boiling water. Pour into a glass with a silver spoon in it. (The spoon will keep the glass from cracking.) Lay a pencil that has a clean cotton string tied to it across the jar. Weight the string with a screw and hang it in the solution. Set the jar in a warm place and let it cool slowly until crystals appear.
3. Dissolve 1 oz. table salt in ½ pt. boiling water and pour it into a shallow plate. Place a string over the edge of the plate. Allow the solution to evaporate slowly for several days and examine the crystals along the string with a magnifying glass. How do these crystals compare with the ones in #1 and #2? Record your observations on the chart.

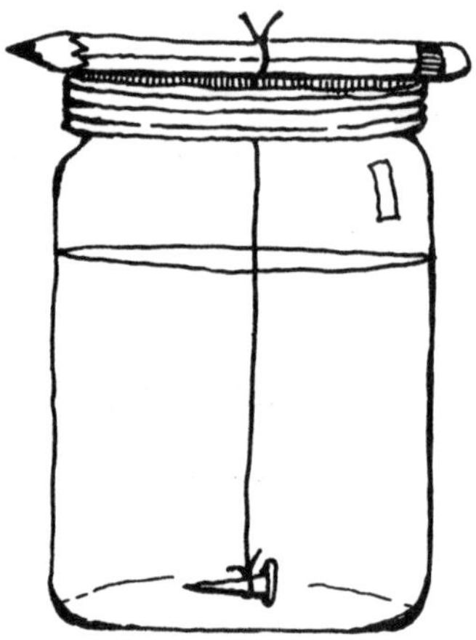

2.

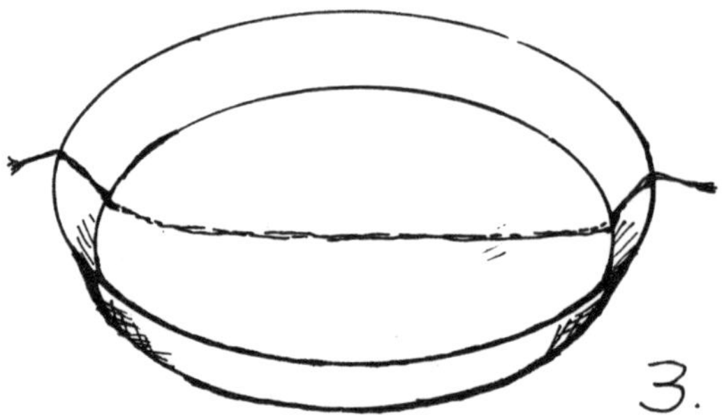

3.

Name ____________________

GROWING CRYSTALS

Record your observations on the chart.

	Day 1	Day 2	Day 3	Day 4	Day 5
1. Salt Crystals (Jar)					
2. Sugar Crystals					
3. Salt Crystals (Dish)					

QUESTIONS:

1. How does the rate of cooling affect crystal growth? Explain.

____

____

____

2. How does limiting the space affect crystal formation? Explain.

____

____

____

3. How does the concentration of salt or sugar in a solution affect the size and shape of crystals? Explain.

____

____

____

Name ______________________

CHARCOAL GARDENS

MATERIALS: Charcoal, Laundry bluing, Food coloring, Salt, Water, Ammonia

DIRECTIONS:

1. Place charcoal in a bowl
2. Mix the following into a solution:
   ¼ c. water
   ¼ c. salt
   ¼ c. laundry bluing
   1 tsp. ammonia
3. Pour the solution over the pieces of charcoal in the bowl trying to spread the solution evenly over the coals. Place some food coloring here and there over the coals. The coals should be about half submerged in the solution.
4. Observe the coals for several days. Describe what happens each day on the "Charcoal Gardens" chart.

Things to Try!

CHANGE THE CONTAINER

Make a charcoal garden in a glass bowl and make another one in an aluminum tray. Observe to see if there is any difference. Record your observations on the chart.

CHANGE THE CHARCOAL

to coal, brick, sponge. Stick toothpicks that have been soaked in food coloring into the sponge. Record your observations on the chart.

CHANGE THE TEMPERATURE

Grow your crystal garden in the refrigerator, over a heater, in a closet, and in the oven. Do you notice any differences? Record your observations on the following chart, "Charcoal Gardens."

Think of an experiment you can do to see if humidity will make any difference in how your crystal garden grows! Write down your experiment. Be certain to include materials needed and the directions.

Can you use sugar instead of salt to grow a crystal garden? ______________

Name ______________________________

CHARCOAL GARDENS

Record your observations each day.

		Day 1	Day 2	Day 3	Day 4	Day 5
CONTAINER	Glass Bowl					
	Aluminum Tray					
MATERIAL	Coal Garden					
	Brick Garden					
	Sponge Garden					
TEMPERATURE	Room Temperature					
	Refrigerator					
	Oven or Heater					

QUESTIONS:

1. Did changing the container have any effect on your garden? Explain.

____

____

2. Did changing the charcoal to another material change the garden? Explain.

____

____

3. Does the temperature affect the growth of your crystal garden? Explain.

____

____

Energy, Force, and Motion

CHAPTER 6

TEACHER NOTES

ENERGY

Energy is the capacity for doing work and exists in many forms, such as heat energy, chemical energy, electrical energy, and mechanical energy.

> Potential energy is stored energy or the ability to do work.
>
> Kinetic energy is the energy of motion.
>
> *A stick of dynamite has potential energy. When heat is applied, the potential energy is converted into kinetic energy with great force.

Work = force acting through a distance or F x D.

Energy cannot be created or destroyed but it can be changed from one form to another. At each energy conversion, some energy is "lost" or converted to a form we cannot use. The total amount of energy never changes.

Water Thermometer
Page 94

EXPANSION AND CONTRACTION IN HEAT ENERGY

Heat energy is the energy of moving molecules and atoms. When something is heated, the molecules move faster. Solids and liquids expand when heated because their molecules move faster and are further apart. When they cool, they contract because the molecules slow down and move closer together. Gases expand more than solids and liquids. If they are in a rigid container, their pressure goes up but they do not expand.

> DEMONSTRATION: How are molecules affected by heat?
>
> 1. Have several students stand together and make a border around them on the floor with tape. Make the border several inches from the people on the outside.
> 2. Ask students to move around slowly. They should stay inside the tape border.
> 3. Keep speeding up the movement in stages until students are no longer able to stay inside the border.
>
> *Discuss how this relates to the movement of molecules when heat is applied.

Heat Exchange Conduction Pages 95 and 96

Conduction, convection, and radiation are the three common ways that energy is transmitted.

Conduction–In conduction, energy is transmitted from atom to atom by direct contact, as in heat conduction when atoms heat up and start vibrating. The vibration spreads from one atom to the next until the whole substance is heated. Conduction occurs next until the whole substance is heated. Conduction occurs in solids, liquids, and gases. It cannot occur in a vacuum. Conductors are materials that carry heat, such as copper and aluminum.

Convection–Convection refers to the transmission of heat energy in liquids and gases. Energy travels from one place to another because of the difference in density of liquids at different temperatures. For example, hot air rises because it is less dense than cold air.

Convection requires a physical medium through which to travel and it does not occur in solids.

Radiation–Radiation is the only form of energy transmission that does not require a physical medium and can travel through a vacuum. Radiant energy is transmitted through radiation. All forms of radiant energy are part of the electromagnetic spectrum which includes light, radio waves, microwaves, ultraviolet light, X-rays, and gamma rays.

Radiation involves changes in the state of electrons. These changes produce electromagnetic waves. The waves are completely undetectable until they come in contact with an object. Each form of electromagnetic radiation interacts with physical objects in a different way.

The activities on these activity sheets deal with heat energy and its transmission by conduction. In the "Heat Exchange" activity sheet, hot and cold metal is added to water to see what effect each has on the water temperature.

Insulators Pages 98 and 99

Using the activity on "Which Materials Are Good Conductors of Heat?" students will test a variety of materials to see which is the best conductor of heat.

Insulators do not carry heat but trap it. Air, polystyrene, and wool are good insulators. These activity sheets include charts for recording water temperature in jars which are insulated with a variety of different materials. Students will determine what is the best insulator and use it to see how long they can keep an ice cube from melting. Have a contest and let students choose different materials to use as insulators!

Ball Drop, Inertia Pages 99 and 100

NEWTON'S THREE LAWS OF MOTION

LAW 1–An object will start to move only if a force is applied to it. Once it is moving, it will continue to move at the same speed in the same direction unless a force is applied. This is sometimes called the law of inertia. Inertia is the resistance to change.

LAW 2–The amount that an object slows down or speeds up (acceleration) varies with the size of the force and the mass of the object. The larger the force and the lighter the object, the greater the acceleration.

LAW 3–For every action, there is an equal and opposite reaction. Whenever anything moves, there is a balance of forces pushing in the opposite direction. When you walk your foot pushes down but the ground pushes up with the same force. Otherwise, your foot would sink into the ground.

GRAVITY

Gravity is a force of attraction that pulls matter together. The strength of the pull depends on the mass of the object. The force of gravity pulls things to earth at the same rate regardless of the objects' sizes or weights. Air resistance is the only thing that will affect the rate of fall or slow it down. In a vacuum, a rock and a feather will fall at the same speed.

The center of gravity is the balance point of an object. In round objects, the center of gravity is the center of the object.

> Making a MOBILE is an excellent activity to apply the concept of "center of gravity."

Activities on "Ball Drop" activity sheet experiment with falling objects and illustrate the force of gravity. Objects are pulled to earth at the same rate regardless of size or weight.

"Inertia" contains activities that help illustrate Newton's first law.

Fast Ball, Friction Fighters Pages 101 and 102

Friction is the force between two things rubbing together. Friction can make things hot. When a moving object is slowed down by friction, its kinetic energy (energy of motion) is changed into heat.

Friction can start things moving, slow things down, or stop things. The roughness of the surface and weight of the object both affect the amount of friction present.

Lubricants reduce friction by filling in rough areas and smoothing out the surface on which an object moves.

Sometimes friction is needed to start things moving if a surface is too smooth. You cannot push off if a surface is too slippery. Treads on tires and soles on shoes provide friction needed to start movement.

The "Fast Ball" activity sheet explores what objects (spheres, discs, or rings) will roll down an inclined plane the fastest.

"Friction Fighters" involves students in an activity to determine which lubricants work the best! Have students try some of these lubricants in the "Fast Ball" activities.

Using a lubricant, is there a significant difference in speed?

What Is Your Horsepower? Page 103

A horse can do 550 foot-pounds of work in one second. This formula is used to determine the horsepower of a car engine:

$$\text{Horsepower} = \frac{\text{Foot Pounds}}{\text{Seconds} \times 550}$$

On this activity sheet, students can calculate their own horsepower by using the same formula. Their horsepower can be compared to the horsepower of a motorbike, a car, and a friend!

Name ______________________________

WATER THERMOMETER

(Heating and Cooling Liquids)

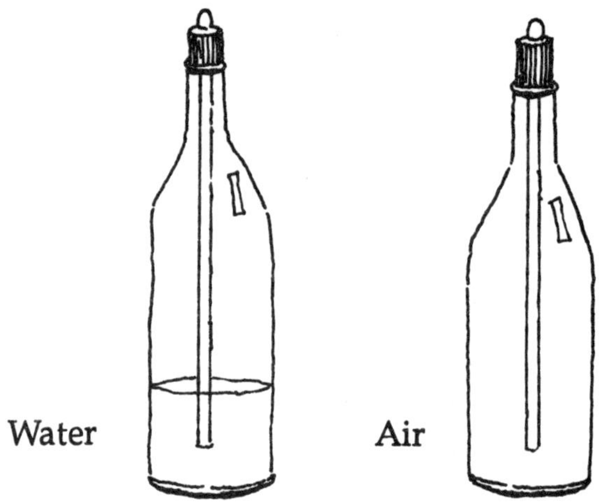

MATERIALS: Water
Narrow-necked jar
One-hole rubber stopper
12 in. clear plastic tubing
Sealing wax
Marker
Food coloring
Masking tape

DIRECTIONS:

1. Fill the jar with cold water. Add food coloring.
2. Insert plastic tubing through the hole in the stopper. There should be no air space below the stopper and water should be forced halfway up the tube. Use sealing wax around the tubing and stopper.
3. Mark the water level in the tube with a marker.
4. Place the jar in sunlight. Check the water level in the tube every 15 minutes for two hours. Mark the level with a marker each time you check it.
5. At the end of two hours, place the jar in a cool place and check the water level every 15 minutes again. Mark the water level with tape each time you check it.

QUESTIONS:

1. What happened to the water level in the tube as the water was warmed? Can you explain why?

   ____

   ____

2. Describe what happened to the water level as the water cooled. Can you explain why?

   ____

3. How could you use this setup as a thermometer? Explain.

   ____

Try This!

Do the same experiment with air, placing a small amount of water in the bottom of the jar and adding food coloring. The tube should be in the water. Explain what happens.

____

Name ______________________________

HEAT EXCHANGE

MATERIALS: Beaker filled with water
Burner
Thermometer
Metal nuts
Tongs

DIRECTIONS:

1. Pour water into the beaker and let it sit for an hour.
   Measure the temperature. ______________________________
2. Using tongs, heat a metal nut in the flame. Then lower it gently into the water.
3. Record the water temperature. ______________________________
4. Add more than one hot nut to the water. What happens?
5. Leave nuts in the freezer overnight. What effect do they have on the water temperature?

Record the results for #4 and #5 on the chart below.

	Water Temperature
0 Nuts	
1 Nut	
2 Nuts	
3 Nuts	
0 Nuts	
1 Frozen Nut	
2 Frozen Nuts	
3 Frozen Nuts	

What conclusion can you make about how hot or cold metal affects water temperature? Write a statement.

____

____

Name ______________________________

WHICH MATERIALS ARE GOOD CONDUCTORS OF HEAT?

MATERIALS: Glass beaker
Beads
Butter
Hot water
Wooden spoon
Metal spoon
Plastic spoon
Drinking straw

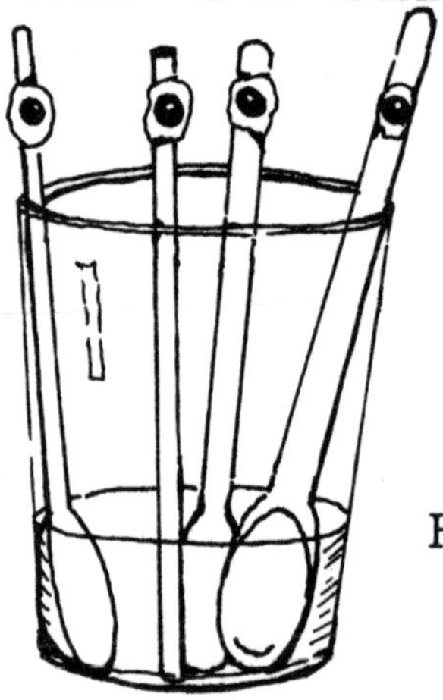

DIRECTIONS:

1. Stick one small bead to each of the test items with a small piece of butter. All of the pieces of butter should be the same size and the bead should be the same distance from the end of each utensil.
2. Stand all of the test items upright in the glass beaker and pour about 3 inches of hot water into the beaker.
3. Heat from the water will be conducted upward through the utensil. The butter will melt and the bead will fall. The first bead to fall shows which material is the best conductor. Explain why.

____

____

____

4. Make a list of the materials you tested (plastic, metal, wood, paper) in order, placing the best conductor of heat first. Then write a concluding statement about the effectiveness of the heat conductors you tested.

1.______________________ 2.______________________

3.______________________ 4.______________________

CONCLUSION:

____

____

Try testing some other materials! Test some different types of metal!

Name ______________________________

INSULATORS

MATERIALS: 2 thermometers
2 small glass jars
Warm/hot water
Towel
Sand
Feathers
Cotton
Newspaper
Polystyrene
Wool

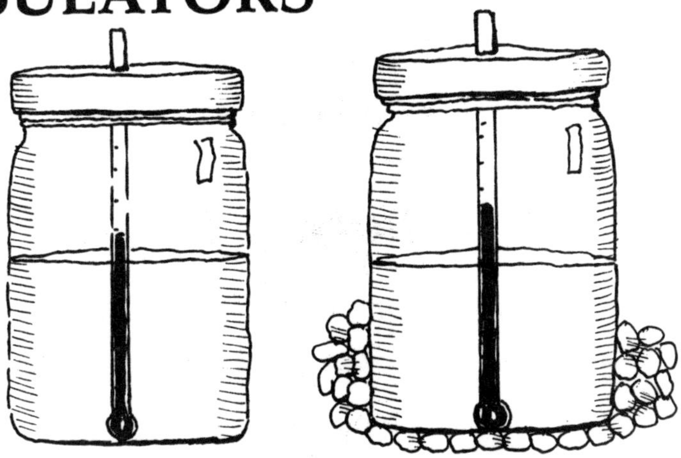

Polystyrene

DIRECTIONS:

1. Make a hole in the top of two jars large enough to insert a thermometer. Seal around the edges of the thermometer with sealing wax or other material.
2. Fill the jars half way with hot water and record the temperature.
3. Wrap one jar with one of the materials above and leave the other jar unwrapped.
4. Take the water temperature in each jar every two minutes. Leave the thermometer in the water a few minutes to be certain you get an accurate reading. Record the results on the chart.
5. Wrap the jar with each of the materials on your list but always start with the water temperature the same in both jars.
6. Try testing some other materials that are not on this list!
7. Make a list rating the insulators you tested. Start with the best insulator and end with the least effective insulator.

	Jar With No Insulation	**Jar With Polystyrene**
Beginning Temperature		
2 Minutes		
4 Minutes		
6 Minutes		
8 Minutes		

Name ______________________

INSULATORS

	Jar With No Insulation	Jar With ________
Beginning Temperature		
2 Minutes		
4 Minutes		
6 Minutes		
8 Minutes		

	Jar With No Insulation	Jar With ________
Beginning Temperature		
2 Minutes		
4 Minutes		
6 Minutes		
8 Minutes		

	Jar With No Insulation	Jar With ________
Beginning Temperature		
2 Minutes		
4 Minutes		
6 Minutes		
8 Minutes		

Name ______________________________

BALL DROP
(falling objects)

MATERIALS: Scrap paper
Tennis ball
Ping-Pong ball
Golf ball

DIRECTIONS:

1. Arrange the balls from heaviest to lightest.
2. If you dropped the heaviest and the lightest balls at the same time, which would hit the floor first?

   Your prediction ______________________________

   What happened ______________________________
3. Test all of the balls. Which hit the floor first?

   ____

Write a rule for falling objects.

____

____

4. Make a series of paper balls that are packed tightly and are the same size as a tennis ball, Ping-Pong ball, and a golf ball.

- Drop the tennis ball and the paper ball of same size. Which hits the floor first?

  ____
- Drop the golf ball and the paper ball of the same size. Which hits the floor first?

  ____
- Drop the Ping-Pong ball and the paper ball of the same size. Which hits the floor first?

  ____

Write a rule that tells what happens when you drop two balls.

____

____

____

Name ______________________________

INERTIA

MATERIALS: 4 or 5 coins the same size
Index card
Sandpaper
Small jar

DIRECTIONS:

1. Put the index card over the mouth of the jar and place a coin on top of the card. The coin should be over the center of the jar. Flick the card forward with your finger so the card moves out from under the coin. The coin should fall into the jar.
2. Place sandpaper over the jar, rough side up. Repeat the directions for #1. Does the coin fall into the jar?
3. Put all but one of the coins in a stack, one on top of the other. Using your forefinger, flick a coin sharply against the bottom coin. The bottom coin should move out leaving the other coins in a stack.

QUESTIONS:

1. Why did the cardboard move in #1?

____

____

2. Why did the coin fall into the jar?

____

3. What happened when you used sandpaper? Can you explain why?

____

____

4. In #3, why didn't the other coins in the stack move forward?

____

____

____

What is *inertia*?

____

____

Name ______________________________

FAST BALL

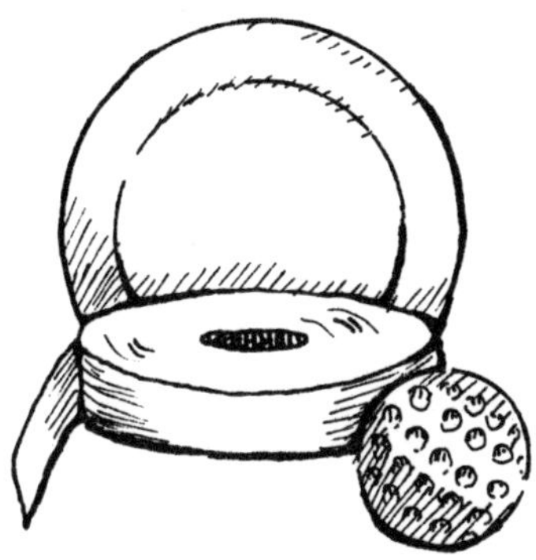

MATERIALS: Inclined planes
Spheres: marbles, golf balls, ball bearings
Disks: coasters, small plastic plates, jar lids
Rings: ring, tire, hula hoop, tape roll, napkin ring

DIRECTIONS:

1. Choose an object from each group.
2. Let all three objects roll down an inclined plane at the same time when someone says "Go" or time each object.
3. Record which object comes in first, second, or third.
4. Repeat this procedure several times until you have used most of the objects above. Record your results on the chart.

	1	2	3
Sphere			
Ring			
Disk			

Write a conclusion about which type of object will roll down an inclined plane the fastest. Can you explain why?

____

____

____

____

____

____

Name ______________________________

FRICTION FIGHTERS
(Lubricants)

MATERIALS:
4 envelopes of unflavored gelatin
Square pan
Bowl
Liquid dish detergent
Vegetable oil
2 bowls
Timer
Table knife
8-oz. cup

	No. of Cubes Transferred
Plain	
Water	
Detergent	
Oil	

DIRECTIONS:

1. Dissolve gelatin in 2 cups hot water.
2. Rub inside of the pan with vegetable oil and pour in the gelatin mixture. Cool until firm (several hours). Cut gelatin into 1" x 1" x 1" cubes.
3. When a signal is given, see how many gelatin cubes you can pick up with two fingers and transfer to another bowl. Continue for 15 seconds.
4. Now pour one-fourth cup of dish detergent over some of the cubes. Make sure cubes are well coated.
5. Repeat #4 and see how many cubes can be transferred in 15 seconds.
6. Put some new cubes in another bowl and pour one-fourth cup water over the cubes. Repeat #4 and see how many cubes can be transferred.
7. Put some new cubes into a clean bowl and repeat the procedure using vegetable oil.

QUESTIONS:

1. According to the experiments you tried above, which liquid was the best lubricant? Explain your answer.

   ____

   ____

2. Why are lubricants necessary in machines that have parts which rub together?

   ____

Try some other lubricants! Record your results!

Name ____________________

WHAT IS YOUR HORSEPOWER?

MATERIALS: Stairs
Stopwatch
Paper and pencil

DIRECTIONS:

1. Time how long it takes you to run a distance of 10 feet up the stairs. Record how many seconds it takes.

James Watt discovered that a horse could do 550 foot-pounds of work in one second. The formula below is used to determine the horsepower of car engines. You can use it to determine what your horsepower might be.

$$\text{Horsepower} = \frac{\text{Foot Pounds}}{\text{Seconds} \times 550}$$

2. Multiply your weight by 10 to give you your foot-pounds. Use the formula above to calculate your horsepower. Divide your foot-pounds by the number of seconds it takes you to climb the stairs x 550.

- A small motorbike has __0-50__ horsepower.
- A car has __100-300__ horsepower.
- I have __________ horsepower!! (See my calculations below.)
- My friend has __________ horsepower!! (See my calculations below.)

Answer Key

CHAPTER 1 – ELECTRICITY

Short Circuit & Fuse (page 4)

1. They shut off the current, protecting the house from a fire.
2. The foil will burn faster and sooner.
3. The foil will not burn as well. If it is too thick it will not burn.
4. They do not need to be replaced.

Complete and Incomplete Circuits (page 8)

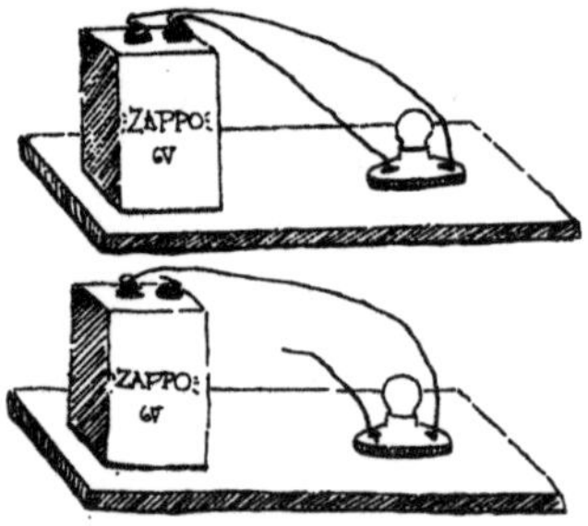

1. A path on which electrons can flow.
2. A complete circuit is a complete path that allows electrons to flow along it. An incomplete circuit is not a complete path and electrons will not flow.
3. Turns the current on and off by opening and closing the circuit.

Predicting Circuits (page 9)

1. Y
2. Y
3. N
4. N
5. N
6. Y
7. N
8. Y

Electric Lingo (page 12)

1. 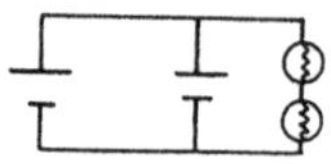

2. 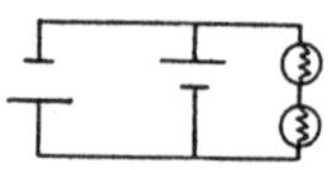

3. 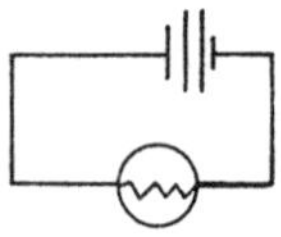

4. 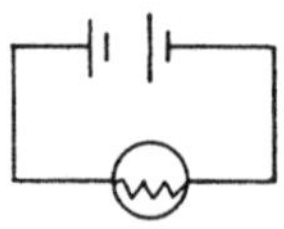

5. 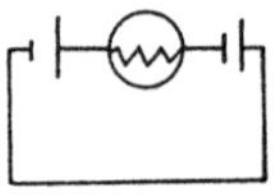

Parallel and Series Circuits (page 13)

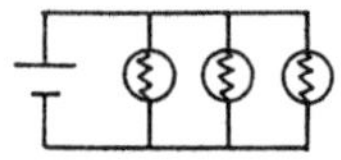

1. The brightness of each bulb stays the same because each bulb is on its own circuit.
2. If one bulb is removed, the other bulbs will stay on because each bulb is on its own circuit.

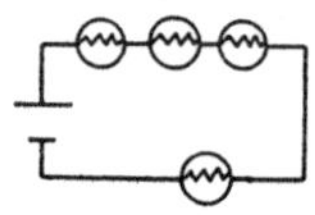

1. As more bulbs are added, the bulbs get dimmer. They are all in the same circuit.
2. If one bulb is removed, all of the bulbs will go out because the circuit will be broken.
3. If Christmas tree lights were in series, all the lights would go out if one bulb burned out. If the lights were in parallel, only the burned out light would be affected.

Lets Make a Battery (page 17)

1. The light bulb goes out because the chemical is no longer present to make electrons collect on the zinc.
2. The light bulb goes out because the chemical is no longer present to make electrons collect on the zinc.

CHAPTER 2 – LIGHT

How Light Travels (page 22)

1. Straight line
2. No
3. Yes. Light travels in a straight line unless refracted by a lens or by travelling through a different substance.

   If light is bent by using a lens, it will not travel in a straight line. You will not see the light looking through the first hole if the cards are in a straight line.

Colors of Light (page 23)

½ red + ½ green = brown

½ red + ½ blue = purple (violet)

½ blue + ½ green = turquoise (blue/green)

Reflection on Different Surfaces (page 25)

1. The mirror reflects light the best.
2. Black card reflects the poorest. It absorbs most of the light.
3. Smooth surfaces and light colors reflect light best.

Observing Images in Mirrors **(page 28)**

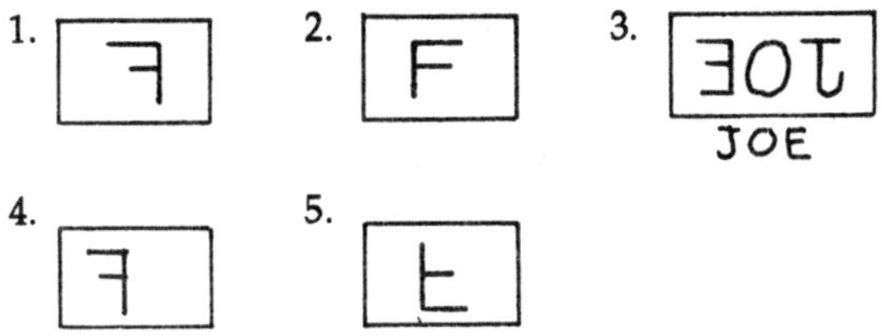

6. Back of spoon – right side up and larger
   Front of spoon – inverted and smaller

1. Images in a flat mirror are flipped from side to side.
2. Images in a curved mirror will be flipped upside down.

- An image in a curved mirror with a flat mirror in front of it will be upside down and backwards.
- OX does not appear upside down because the letters are the same on top and bottom.

Solar Reflector Oven **(page 30)**

Light hitting the curved surface covered with foil will be reflected and refracted into the box, heating the inside and melting the marshmallows.

Refraction **(page 31)**

The Bent Pencil –
Pencil appears bent because light is refracted as it passes from air to water.

The Magic Penny –
Penny appears to be in another place because the light is refracted when it hits the water.

Can Liquids Magnify? **(page 32)**

Different liquids will magnify objects because the light rays are refracted as they pass through another substance. The surface of the liquid is also curved which makes the object appear larger.

Shadows **(page 34)**

1. If you halve the straw length, you halve the shadow length.
2. If you quarter the straw length, you quarter the shadow.
3. The length of the straw is in direct proportion to the length of the shadow.
4. Brightness does not affect shadow length.
5. Distance between light and straw does not affect shadow length.

CHAPTER 3 – SOUND

Jingling Jar **(page 41)**

When hot water was added to the jar, some of the molecules escaped from the jar because the air expanded. Because there were fewer molecules in the jar when the lid was put back, the bell could not be heard as clearly the second time.

Bottle Band **(page 42)**

Blowing across the top of the bottle makes the air in the bottle vibrate.

1. The lowest sound is made by the bottle with the most air, or least water.
2. The highest sound is made by the bottle with the least air, or most water.
3. The more air there is to vibrate, the lower the sound.

Shoe Box Guitar **(page 43)**

1. When the line is pulled tighter, the sound is higher.
2. Cutting a hole in the box will increase the resonance or intensity of the sound.
3. The pitch is higher as the speed of vibration increases.

Echoes **(page 44)**

Activity I

1. Yes.
2. You hear an echo when you begin walking until you get near the end of the hall. At the end of the hall the sound changes.
3. Yes. When passing open doors, the sound is not so loud. Some of the vibrations enter the room and are absorbed.

Activity II

1. Construction of booths and curtains used in auditoriums absorb some of the sound. The booth can be soundproof from inside. Curtains prevent the sound from bouncing off the walls in an auditorium.

CHAPTER 4 – SIMPLE MACHINES

Lifting Loads With a First-Class Lever **(page 51)**

1. Load is easier to lift when it is closer to the fulcrum.
2. The effort arm is longer.
3. The longer the effort arm, the less force is needed to lift the load.

Relationship of Force and Distance First-Class Lever **(page 53)**

1. The distance the effort arm travels increases.
2. The distance the load arm travels decreases.
3. As the fulcrum gets closer to the load, the effort to lift the load decreases. The effort arm travels a greater distance and the load travels a lesser distance. Distance is sacrificed.

Second-Class Lever **(page 55)**

1. It is easier to lift books when they are closer to the fulcrum.
2. It is easier when the effort arm is longer.
3. It is easier when the effort arm is longer.

Third-Class Lever **(page 56)**

1. It is easier to lift books when the hand is farther from fulcrum. When the hand is farther away from the effort arm, distance is greater.

2. With first and second-class levers, the load is easier to lift near the fulcrum. In a third-class lever, with the effort between the fulcrum and the load, the effort is greater than the load, but the load travels faster and farther. The advantage with a third-class lever is speed and distance.

Wheel and Axle **(page 57)**

1. A wheel and axle helps something turn more easily – you use less force.

   If the wheel is turning the axle, it is a second-class lever and there is increased force but less speed and distance. (Door knob, screwdriver, steering wheel)

   If axle is turning the wheel, it is a third-class lever with a gain in speed and distance but a decrease in force.
2. Roller skates, car wheels, wheels in machines (see #1)

As the Thread Turns **(page 61)**

1. A smaller depth is drilled with the more threaded screw.
2. The more threads, the more turns, but the easier it is to turn because it takes less effort for each turn.

Fixed Pulley **(page 62)**

1. Using the pulley does not decrease the force.
2. Using the pulley makes the books move in the opposite direction from the force.
3. A fixed pulley does not alter the amount of force used to lift an object but it reverses the direction of the force. When you pull down the load moves up.

Movable Pulley **(page 63)**

1. Yes. The pulley decreases the amount of force needed to lift the books.
2. The load moves in the same direction as the pulley.
3. When using a movable pulley the load travels in the same direction as the effort. However, the amount of force required to lift the load is less than the weight of the load.

Combining Fixed and Movable Pulleys **(page 64)**

1. The amount of force needed to lift the load is decreased.
2. The load moves in the opposite direction from the effort.
3. The combined pulley system offers the advantage of both. The load can be moved with half as much force (because of the movable pulley) and the load can be moved upward by pulling downward.

CHAPTER 5 – MATTER

Making Carbon Dioxide **(page 71)**

5. When vinegar and bicarbonate of soda are mixed CO_2 is produced. Gas molecules take up more space than those of a solid or liquid, and inflate the balloon.

Molecules Meshing Together **(page 75)**

Amount of liquid is less than two cups. Some of the alcohol molecules fill in the spaces between the water molecules.

Mysterious Mixture:

Iron filings can be separated by using the magnet, or the mixture can be put in warm water. The sugar will dissolve, leaving the iron filings. The filings can be strained from the water.

"Goop" **(page 76)**

Students can feel the change when a cross-linking ion (sodium tetraborate—Borax and water) is mixed with a polymer (polyvinyl acetate – white glue).

Miscible and Immiscible Liquids **(page 85)**

1. The oil will form a layer on top of the water because water and oil do not dissolve in each other.
2. The food coloring will stay in little balls in the oil because it will not dissolve.
3. When the food coloring hits the water, it will begin to dissolve and color the water. The balls will disappear.

Growing Crystals Chart **(page 87)**

1. Slow cooling forms larger crystals.
2. Will not change shape unless it hits side of container—then crystal will be deformed in some way.
3. The solution must be saturated for crystals to form.

CHAPTER 6 – FORCE, ENERGY, AND MOTION

Water Thermometer **(page 94)**

1. As the water heats, the water level rises because the molecules begin to move faster and farther apart. This causes the water to expand.
2. As the water cools, the molecules slow down and move closer together. The water level in the tube falls.

Ball Drop **(page 99)**

Rule for falling objects—the force of gravity pulls things to earth at the same rate regardless of size or weight.

Inertia **(page 100)**

1. Force was applied to the cardboard so it moved forward.
2. No force was applied to the penny, so it dropped into the jar.
3. Sandpaper has a rough surface causing friction so it does not move forward as well as the cardboard.
4. Only the bottom coin moved because it was the only one that had force applied to it.

   Inertia is the resistance to change.

Fast Ball **(page 101)**

A sphere will roll down an inclined plane faster than the other objects, because of its center of gravity which is in the middle of the object.